Electronics Engineering: Principles and Applications

Electronics Engineering: Principles and Applications

Hannah Clark

MURPHY & MOORE
www.murphy-moorepublishing.com

Published by Murphy & Moore Publishing
1 Rockefeller Plaza,
New York City, NY 10020, USA
www.murphy-moorepublishing.com

Electronics Engineering: Principles and Applications
Hannah Clark

International Standard Book Number: 978-1-63987-184-1 (Hardback)

Cataloging-in-Publication Data

Electronics engineering : principles and applications / Hannah Clark.
p. cm.
Includes bibliographical references and index.
ISBN 978-1-63987-184-1
1. Electronics. 2. Electrical engineering. 3. Engineering. I. Clark, Hannah.
TK7815 .E44 2022
621.381--dc23

Contents

Preface

Electronics engineering is a sub-discipline of electrical engineering which makes use of nonlinear and active electrical devices like transistors and diodes for designing electronic circuits and systems. Integrated circuits and printed circuit boards are also important parts of this discipline. Electronics engineering can be further classified into various sub-fields such as solid state physics, telecommunications engineering, signal processing, systems engineering, robotics, VLSI design and instrumentation engineering. Electronic circuits can be divided into analog and digital circuits. Analog circuits include amplifiers, oscillators, function generators, and wave shaping circuits. Multiplexers, decoders and microprocessors are some prominent examples of digital circuits. Electronics engineering finds extensive applications across various fields such as consumer electronics, industrial automation and aerospace industry. Some of the emerging areas of research under this field are image processing, motion control and smart grid systems. This book unfolds the innovative aspects of electronics engineering which will be crucial for the holistic understanding of the subject matter. Some of the diverse topics covered herein address the varied branches that fall under this category. Those in search of information to further their knowledge will be greatly assisted by this book.

A foreword of all chapters of the book is provided below:

Chapter 1 - Electronic engineering is a specialization within the domain of engineering associated with the use of active electronic components and electromagnetic spectrum to design integrated circuits, electronic circuits, devices and their systems. Electronics deals with the control of emission and flow of electrons in vacuum and matter by amplification and rectification. This is an introductory chapter which will introduce briefly all the significant aspects of electronic engineering; **Chapter 2** - Digital electronics is a branch of electronics which studies circuits that operate on digital signals. Digital circuits accept and process binary data according to boolean logic. In order to completely understand digital electronics and digital circuits, it is necessary to understand the processes related to it. This chapter elucidates the varied processes and mechanisms associated with this area of study; **Chapter 3** - Power electronics deals with electronic devices and associated components for the conversion and control of electric power. It makes the use of semiconductor devices like diodes, power transistors and thyristors for conversion. This chapter closely examines the key concepts of power electronics to provide an extensive understanding of the subject; **Chapter 4** - An electrical circuit is a closed loop which provides a path for the flow of electrons. There are three basic circuit elements which are necessary for the electric current to flow safely through an electrical circuit, namely, a source of energy connecting wires and devices which use current. All the diverse aspects of electrical circuits such as the circuit diagram, phasor diagram, electrical impedance, Ohm's law, etc. have been carefully analyzed in this chapter; **Chapter 5** - An assembly of electronic circuits on a thin substrate of semiconductor is known as an integrated circuit. The logic and techniques required to design integrated circuits fall under integrated circuit design. This chapter discusses in detail the theories and concepts related to integrated circuit ad integrated circuit design; **Chapter 6** - Electronics are applied in various areas like audio electronics, radio electronics, consumer electronics, nanoelectronics, molecular electronics, bioelectronics and nuclear electronics. The diverse applications of electronics in these areas have been thoroughly discussed in this chapter.

At the end, I would like to thank all the people associated with this book devoting their precious time and providing their valuable contributions to this book. I would also like to express my gratitude to my fellow colleagues who encouraged me throughout the process.

Hannah Clark

1

An Introduction to Electronic Engineering

Electronic engineering is a specialization within the domain of engineering associated with the use of active electronic components and electromagnetic spectrum to design integrated circuits, electronic circuits, devices and their systems. Electronics deals with the control of emission and flow of electrons in vacuum and matter by amplification and rectification. This is an introductory chapter which will introduce briefly all the significant aspects of electronic engineering.

Electronics

Electronics is the branch of physics and electrical engineering that deals with the emission, behavior, and effects of electrons and with electronic devices. Electronics encompasses an exceptionally broad range of technology. The term originally was applied to the study of electron behavior and movement, particularly as observed in the first electron tubes. It came to be used in its broader sense with advances in knowledge about the fundamental nature of electrons and about the way in which the motion of these particles could be utilized. Today many scientific and technical disciplines deal with different aspects of electronics. Research in these fields has led to the development of such key devices as transistors, integrated circuits, lasers, and optical fibers. These in turn have made it possible to manufacture a wide array of electronic consumer, industrial, and military products. Indeed, it can be said that the world is in the midst of an electronic revolution at least as significant as the industrial revolution of the 19th century.

The Vacuum Tube Era

Theoretical and experimental studies of electricity during the 18th and 19th centuries led to the development of the first electrical machines and the beginning of the widespread use of electricity. The history of electronics began to evolve separately from that of electricity late in the 19th century with the identification of the electron by the English physicist Sir Joseph John Thomson and the measurement of its electric charge by the American physicist Robert A. Millikan in 1909.

At the time of Thomson's work, the American inventor Thomas A. Edison had observed a bluish glow in some of his early light bulbs under certain conditions and found that a current would flow from one electrode in the lamp to another if the second one (anode) were made positively charged with respect to the first (cathode). Work by Thomson and his students and by the English engineer John Ambrose Fleming revealed that this so-called Edison effect was the result of the emission of electrons from the cathode, the hot filament in the lamp. The motion of the electrons to the anode, a metal plate, constituted an electric current that would not exist if the anode were negatively charged.

This discovery provided impetus for the development of electron tubes, including an improved X-ray tube by the American engineer William D. Coolidge and Fleming's thermionic valve (a two-electrode vacuum tube) for use in radio receivers. The detection of a radio signal, which is a very high-frequency alternating current (AC), requires that the signal be rectified; i.e., the alternating current must be converted into a direct current (DC) by a device that conducts only when the signal has one polarity but not when it has the other—precisely what Fleming's valve (patented in 1904) did. Previously, radio signals were detected by various empirically developed devices such as the "cat whisker" detector, which was composed of a fine wire (the whisker) in delicate contact with the surface of a natural crystal of lead sulfide (galena) or some other semiconductor material. These devices were undependable, lacked sufficient sensitivity, and required constant adjustment of the whisker-to-crystal contact to produce the desired result. Yet these were the forerunners of today's solid-state devices. The fact that crystal rectifiers worked at all encouraged scientists to continue studying them and gradually to obtain the fundamental understanding of the electrical properties of semiconducting materials necessary to permit the invention of the transistor.

In 1906 Lee De Forest, an American engineer developed a type of vacuum tube that was capable of amplifying radio signals. De Forest added a grid of fine wire between the cathode and anode of the two-electrode thermionic valve constructed by Fleming. The new device, which De Forest dubbed the Audion, was thus a three-electrode vacuum tube. In operation, the anode in such a vacuum tube is given a positive potential (positively biased) with respect to the cathode, while the grid is negatively biased. A large negative bias on the grid prevents any electrons emitted from the cathode from reaching the anode; however, because the grid is largely open space, a less negative bias permits some electrons to pass through it and reach the anode. Small variations in the grid potential can thus control large amounts of anode current.

The vacuum tube permitted the development of radio broadcasting, long-distance telephone, television, and the first electronic digital computers. These early electronic computers were, in fact, the largest vacuum-tube systems ever built.

The special requirements of the many different applications of vacuum tubes led to numerous improvements, enabling them to handle large amounts of power, operate at very high frequencies, have greater than average reliability, or be made very compact (the size of a thimble). The cathode-ray tube, originally developed for displaying electrical waveforms on a screen for engineering measurements, evolved into the television picture tube. Such tubes operate by forming the electrons emitted from the cathode into a thin beam that impinges on a fluorescent screen at the end of the tube. The screen emits light that can be viewed from outside the tube. Deflecting the electron beam causes patterns of light to be produced on the screen, creating the desired optical images.

Notwithstanding the remarkable success of solid-state devices in most electronic applications, there are certain specialized functions that only vacuum tubes can perform. These usually involve operation at extremes of power or frequency.

Vacuum tubes are fragile and ultimately wear out in service. Failure occurs in normal usage either from the effects of repeated heating and cooling as equipment is switched on and off (thermal fatigue), which ultimately causes a physical fracture in some part of the interior structure of the tube, or from degradation of the properties of the cathode by residual gases in the tube. Vacuum tubes also take time (from a few seconds to several minutes) to "warm up" to operating temperature—an inconvenience at best and in some cases a serious limitation to their use. These shortcomings motivated scientists to seek an alternative to the vacuum tube and led to the development of the transistor.

The Semiconductor Revolution

Invention of the Transistor

The invention of the transistor in 1947 by John Bardeen, Walter H. Brattain, and William B. Shockley provided the first of a series of new devices with remarkable potential for expanding the utility of electronic equipment. Transistors, along with such subsequent developments as integrated circuits, are made of crystalline solid materials called semiconductors, which have electrical properties that can be varied over an extremely wide range by the addition of minuscule quantities of other elements. The electric current in semiconductors is carried by electrons, which have a negative charge, and also by "holes," analogous entities that carry a positive charge. The availability of two kinds of charge carriers in semiconductors is a valuable property exploited in many electronic devices made of such materials.

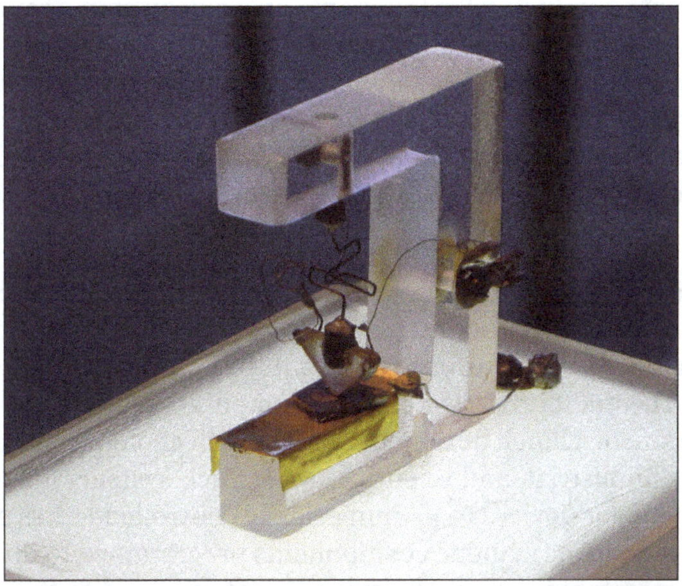

Transistor: The first transistor, invented by American physicists John Bardeen, Walter H. Brattain, and William B. Shockley.

Early transistors were produced using germanium as the semiconductor material, because methods of purifying it to the required degree had been developed during and shortly after World War II. Because the electrical properties of semiconductors are extremely sensitive to the slightest trace of certain other elements, only about one part per billion of such elements can be tolerated in material to be used for making semiconductor devices.

During the late 1950s, research on the purification of silicon succeeded in producing material suitable for semiconductor devices, and new devices made of silicon were manufactured from about 1960. Silicon quickly became the preferred raw material, because it is much more abundant than germanium and thus less expensive. In addition, silicon retains its semiconducting properties at higher temperatures than does germanium. Silicon diodes can be operated at temperatures up to 200 °C (400 °F), whereas germanium diodes cannot be operated above 85 °C (185 °F). There was one other important property of silicon, not appreciated at the time but crucial to the development of low-cost transistors and integrated circuits: silicon, unlike germanium, forms a tenaciously adhering oxide film with excellent electrical insulating properties when it is heated to high temperatures in the presence of oxygen. This film is utilized as a mask to permit the desired impurities that modify the electrical properties of silicon to be introduced into it during manufacture of semiconductor devices. The mask pattern, formed by a photolithographic process, permits the creation of tiny transistors and other electronic components in the silicon.

Integrated Circuits

By 1960 vacuum tubes were rapidly being supplanted by transistors, because the latter had become less expensive, did not burn out in service, and were much smaller and more reliable. Computers employed hundreds of thousands of transistors each. This fact, together with the need for compact, lightweight electronic missile-guidance systems, led to the invention of the integrated circuit (IC).

Early ICs contained about 10 individual components on a silicon chip 3 mm (0.12 inch) square. By 1970 the number was up to 1,000 on a chip of the same size at no increase in cost. Late in the following year the first microprocessor was introduced. The device contained all the arithmetic, logic, and control circuitry required to perform the functions of a computer's central processing unit (CPU). This type of large-scale IC was developed by a team at Intel Corporation, the same company that also introduced the memory IC in 1971. The stage was now set for the computerization of small electronic equipment.

Until the microprocessor appeared on the scene, computers were essentially discrete pieces of equipment used primarily for data processing and scientific calculations. They ranged in size from minicomputers, comparable in dimensions to a small filing cabinet, to mainframe systems that could fill a large room. The microprocessor enabled computer engineers to develop microcomputers—systems about the size of a lunch box or smaller but with enough computing power to perform many kinds of business, industrial, and scientific tasks. Such systems made it possible to control a host of small instruments or devices (e.g., numerically controlled lathes and one-armed robotic devices for spot welding) by using standard components programmed to do a specific job. The very existence of computer hardware inside such devices is not apparent to the user.

The large demand for microprocessors generated by these initial applications led to high-volume production and a dramatic reduction in cost. This in turn promoted the use of the devices in many other applications—for example, in household appliances and automobiles, for which electronic controls had previously been too expensive to consider. Continued advances in IC technology gave rise to very large-scale integration (VLSI), which substantially increased the circuit density of microprocessors. These technological advances, coupled with further cost reductions stemming from improved manufacturing methods, made feasible the mass production of personal computers for use in offices, schools, and homes.

By the mid-1980s inexpensive microprocessors had stimulated computerization of an enormous variety of consumer products. Common examples included programmable microwave ovens and thermostats, clothes washers and dryers, self-tuning television sets and self-focusing cameras, videocassette recorders and video games, telephones and answering machines, musical instruments, watches, and security systems. Microelectronics also came to the fore in business, industry, government, and other sectors. Microprocessor-based equipment proliferated, ranging from automatic teller machines (ATMs) and point-of-sale terminals in retail stores to automated factory assembly systems and office workstations.

By mid-1986 memory ICs with a capacity of 262,144 bits (binary digits) were available. In fact, Gordon E. Moore, one of the founders of Intel, observed as early as 1965 that the complexity of ICs was approximately doubling every 18–24 months, which was still the case in 2000. This empirical "Moore's law" is widely used in forecasting the technological requirements for manufacturing future ICs.

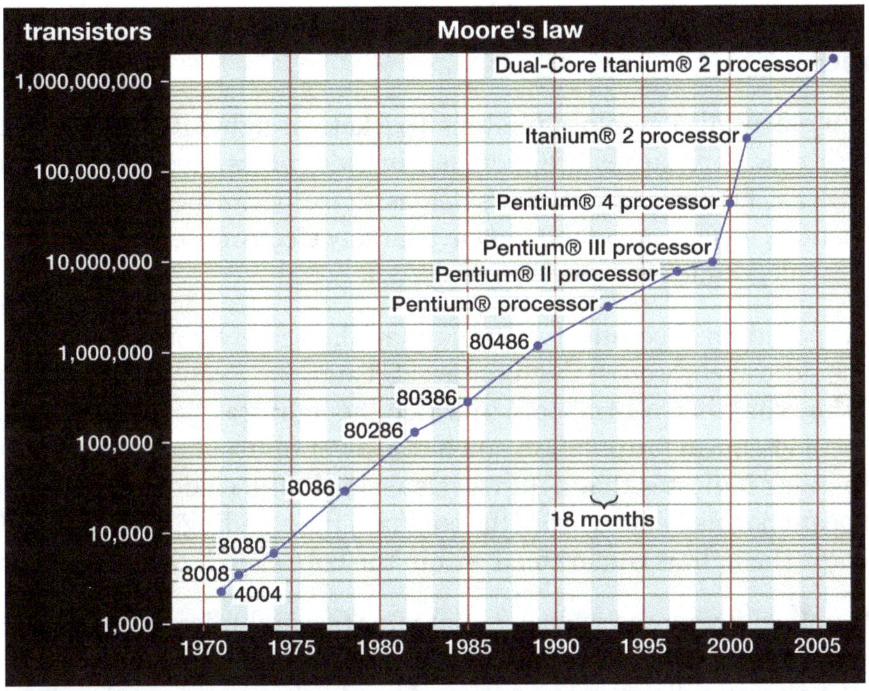

Moore's law: Gordon E. Moore observed that the number of transistors on a computer chip was doubling about every 18–24 months. As shown in the logarithmic graph of the number of transistors on Intel's processors at the time of their introduction, his "law" was being obeyed.

Compound Semiconductor Materials

Many semiconductor materials other than silicon and germanium exist, and they have different useful properties. Silicon carbide is a compound semiconductor, the only one composed of two elements from column IV of the periodic table. It is particularly suited for making devices for specialized high-temperature applications. Other compounds formed by combining elements from column III of the periodic table—such as aluminum, gallium, and indium—with elements from column V—such as phosphorus, arsenic, and antimony—are of particular interest. These so-called III-V compounds are used to make semiconductor devices that emit light efficiently or that operate at exceptionally high frequencies.

A remarkable characteristic of these compounds is that they can, in effect, be mixed together. One can produce gallium arsenide or substitute aluminum for some of the gallium or also substitute phosphorus for some of the arsenic. When this is done, the electrical and optical properties of the material are subtly changed in a continuous fashion in proportion to the amount of aluminum or phosphorus used.

Except for silicon carbide, these compounds have the same crystal structure. This makes possible the gradation of composition, and thus the properties, of the semiconductor material within one continuous crystalline body. Modern material-processing techniques allow these compositional changes to be controlled accurately on an atomic scale.

These characteristics are exploited in making semiconductor lasers that produce light of any given wavelength within a considerable range. Such lasers are used, for example, in compact disc players and as light sources for optical fiber communication.

Optoelectronics

A new direction in electronics employs photons (packets of light) instead of electrons. By common consent these new approaches are included in electronics, because the functions that are performed are, at least for the present, the same as those performed by electronic systems and because these functions usually are embedded in a largely electronic environment. This new direction is called optical electronics or optoelectronics.

In 1966 it was proposed on theoretical grounds that glass fibers could be made with such high purity that light could travel through them for great distances. Such fibers were produced during the early 1970s. They contain a central core in which the light travels. The outer cladding is made of glass of a different chemical formulation and has a lower optical index of refraction. This difference in refractive index indicates that light travels faster in the cladding than it does in the core. Thus, if the light beam begins to move from the core into the cladding, its path is bent so as to move it back into the core. The light is constrained within the core even if the fiber is bent into a circle.

The core of early optical fibers was of such a diameter (several micrometers [μm], or about one-tenth the diameter of a human hair) that the various rays of light in the core could travel in slightly different paths, the shortest directly down the axis and other longer paths wandering back and forth across the core. This limited the maximum distance that a pulse of light could travel without becoming unduly spread by the time it arrived at the receiving end of the fiber, with the central ray arriving first and others later. In a digital communications system, successive pulses can overlap one another and be indistinguishable at the receiving end. Such fibers are called multimode fibers, in reference to the various paths (or modes) that the light can follow.

During the late 1970s, fibers were made with smaller core diameters in which the light was constrained to follow only one path. This occurs if the core has a diameter a little larger than the wavelength of the light traveling in it—i.e., about 10 to 15 μm (0.01 to 0.015 mm, or 0.0004 to 0.0006 inch). These single-mode fibers avoid the difficulty described above. By 1993 optical fibers capable of carrying light signals more than 215 km (135 miles) became available. Such distance records have become obsolete with the use of specialized fibers that incorporate integral amplifying features. Fibers employing these optical amplifiers carry light signals over transoceanic distances without the conventional pulse regeneration measures that were needed in the past.

Optical fibers have several advantages over copper wires or coaxial cables. They can carry information at a much higher rate, they occupy less space (an important feature in large cities and in buildings), and they are quite insensitive to electrical noise. Moreover, it is virtually impossible to make unauthorized connections to them. Costs, initially high, have dropped to the point where most new installations of telephone circuits between switching centers and over longer distances consist of optical fibers.

Given the fact that communication signals arrive at a central switching office in optical form, it has been attractive to consider switching them from one route to another by optical means rather than electrically, as is done today. The distances between central offices in most cases are substantially shorter than the distance light can travel within a fiber. Optical switching would make unnecessary the detection and regeneration of the light signals, steps that are currently required. Such optical central-office switches are ready for installation today and will further advance the dramatic changes wrought by the use of light waves rather than electrons.

Another direction in optoelectronics builds in part on the foregoing developments but to a quite different end. A key problem in developing faster computers and faster integrated circuits to use in them is related to the time required for electrical signals to travel over wire interconnections. This is a difficulty both for the integrated circuits themselves and for the connections between them. Under the best circumstances, electrical signals can travel in a wire at about 90 percent of the speed of light. A more usual rate is 50 percent. Light travels about 30 cm (12 inches) in a billionth of a second. Modern computers operate at speeds of more than one billion operations per second. Thus, if two signals that start simultaneously from different sites are to arrive at their destination simultaneously, the paths they travel must not differ in length by more than a few centimeters.

Two approaches can be envisioned. In one, all the integrated circuits are placed as close together as possible to minimize the distances that signals must travel. This creates a cooling problem, because the integrated circuits generate heat. In the other possible approach, all the paths for signals are made equal to the longest path. This requires the use of much more wire, because most paths are longer than they would otherwise be. All this wire takes space, which means that the integrated circuits have to be placed farther apart than is preferable.

Ultimately, as computers operate even faster, neither approach will work, nor a radically new technique must be used. Optical communication between integrated circuits is one possible answer. Light beams do not take up space or interfere with cooling air. If the communication is optical, then the computation might be done optically as well. Optical computation will require a radically different form of integrated circuit, which can in principle be made of gallium arsenide and related III-V compounds. These matters are currently under serious study in research laboratories.

Superconducting Electronics

Numerous metals completely lose their resistance to the flow of electric current at temperatures approaching absolute zero (0 K, −273 °C, or −460 °F) and become superconducting. Other equally dramatic changes in electrical properties occur as well. One of these is the Josephson Effect, named for the British physicist Brian D. Josephson, who predicted and then discovered the phenomenon in 1962. The Josephson Effect governs the passage of current from one superconducting metal to another through a very thin insulating film between them (the Josephson junction) and the effects of small magnetic fields on this current.

Josephson junction devices change from one electrical state to another in extraordinarily short times, offering the possibility of producing superconducting microcircuits that operate faster than any other kind known. Serious efforts have been made to construct a computer on this basis, but most of the projects have been either discontinued or sharply cut back because of technical difficulties. Interest in the approach has also waned because of increases in the speed of III-V semiconductor microcircuits.

Josephson junctions have other uses in science. They make extremely sensitive detectors of small magnetic fields, for example. The voltage across a Josephson junction is known on theoretical grounds to be dependent only on the values of certain basic physical constants. Since these constants are known to great accuracy, Josephson junctions are now used to provide the absolute standard of voltage.

Other important applications of Josephson junctions have to do with the metrology of very high-speed signals. Measurements of fast phenomena require the use of even faster measurement tools, which Josephson devices provide.

Flat-Panel Displays

Display devices convey information in visible form from electronic devices to human viewers. Common examples are the faces on digital watches, numerical indicators on stereo equipment, and the picture tubes in television sets and computer monitors. Until recently the most versatile of these has been the picture tube, which can present numbers, letters, graphs, and both still and moving pictures. While picture tubes set a very high standard of performance and provide bright colour images, they are bulky, heavy, and expensive. Designers of television receivers have long desired a display device having the virtues of the picture tube but fewer of the disadvantages, so that a "picture on the wall" television set can be produced.

New developments in flat-panel displays have made this possible. Such displays are advanced versions of the liquid crystal display familiar in digital watch faces. They are essentially two parallel sheets of thin glass having the facing sides coated with a transparent yet electrically conducting film such as indium tin oxide. The film layer nearer the viewer is patterned, while the other layer is not. The space between the films is filled with a fluid with unusual electrical and optical properties, so that, if an electrical field is established between the two thin films, the molecules of the fluid line up in such a way that the light-reflecting or light-transmitting properties of the assembly are radically changed. The electro-optical fluid is an electrical insulator, so very little electric current flows. Thus, almost no power is consumed, making the display well suited for use in battery-powered applications. All flat-panel displays have these characteristics in common, but the many different varieties exploit the electro-optical effects in numerous ways.

Displays that produce images are patterned with myriads of tiny picture elements that can be electrically activated independently to produce patterns of light and dark or arbitrary forms. Superposed colour filters having arrays of elements corresponding to those in the display permit the formation of colour images of a quality rivaling that of colour cathode-ray tube displays. Such displays are used as viewing devices for television sets, computers, and video and digital cameras.

Colour displays capable of serving as television screens or computer displays are available in sizes of more than 35 cm (15 inches) on the diagonal, at costs nearly competitive with picture tubes. There is strong demand for them in laptop computers, where the thinness of a flat-panel display is essential. Such displays have more than three million separate elements in the picture array, each of which must have separate means for its control. The control electronics is integrated into the display, for otherwise the number of individual wires needed to connect with the rest of the circuitry would be prohibitive.

A great amount of effort is being expended to increase the size and decrease the cost of flat-panel displays, because the potential market for them is clearly substantial. Much of the reduction in cost is obtained through experience in manufacturing, where low yields attributable to defects in the patterns have been a major problem.

The Science of Electronics

Valence Electrons

Since electronics is concerned with the control of the motion of electrons, one must keep in mind that electrons, being negatively charged, are attracted to positive charges and repelled by other negative charges. Thus, electrons in a vacuum tend to space themselves apart from one another and form a cloud, subject to the influences of other charges that may be present. An electric current is created by the motion of electrons, whether in a vacuum, in a wire, or in any other electrically conducting medium. In each of these cases, electrons move as a result of their attraction to positive charges or repulsion from negative ones.

An atom consists of a nucleus of protons and neutrons around which electrons, equal in number to the protons in the nucleus, travel in orbits much like those of the planets around the Sun. Because of this equality in the number of positively and negatively charged constituent particles, the atom as a whole is electrically uncharged. When atoms are combined into certain solids called covalent solids (notably the elements of column IV of the periodic), the valence electrons (outer electrons) are shared between neighboring atoms, and the atoms thereby become bound together. This occurs not only in elemental solids, wherein all the atoms are of the same kind, but also in chemical compounds (e.g., the III-V compounds).

Different materials vary greatly in their ability to conduct electricity, depending directly on the ease or difficulty of setting electrons free from their atoms. In insulating materials all the outermost electrons of the atoms are tightly bound in the chemical bonds between atoms and are not free to move. In metals there are more valence electrons than are required for bonding, and these excess electrons are freely available for electrical conduction.

Most insulators and metals are crystalline materials but are composed of a great many very small crystals. (In all crystals the atoms are positioned in a regularly spaced three-dimensional array.) Semiconducting solids for electronic applications, however, are prepared as single large crystals. The fact that the atoms in a semiconductor are arranged in a periodic, three-dimensional array of large size (large, that is, in comparison with an atom) makes the atoms appear nearly invisible to electrons moving within a crystal. The reasons for this behavior are too complex to explain here, but this property allows electrons to be quite mobile in semiconductors.

Conduction in Semiconductors

In semiconductors such as silicon (which is used as the example here), each constituent atom has four outer electrons, each of which pairs with an electron from one of four neighboring atoms to form the interatomic bonds. Completely pure silicon thus has essentially no electrons available at room temperature for electronic conduction, making it a very poor conductor. However, if an atom from column V of the periodic, such as phosphorus, is substituted for an atom of silicon, four of its five outer electrons will be used for bonding, while the fifth will be free to move within the crystal. If the replacement atom comes from column III of the periodic table—say, boron—it will have only three outer electrons, one too few to complete the four interatomic bonds. The facts that the crystal would be electrically neutral were this bond complete means that, if an electron is missing, the vacancy will have a positive charge. A neighboring electron can move into the vacancy, leaving another vacancy in the electron's former place. This vacancy, with its positive charge, is thus mobile and is called a "hole". Holes in semiconductors move about as readily as electrons do, but, because they are positively charged, they move in directions opposite to the motion of electrons.

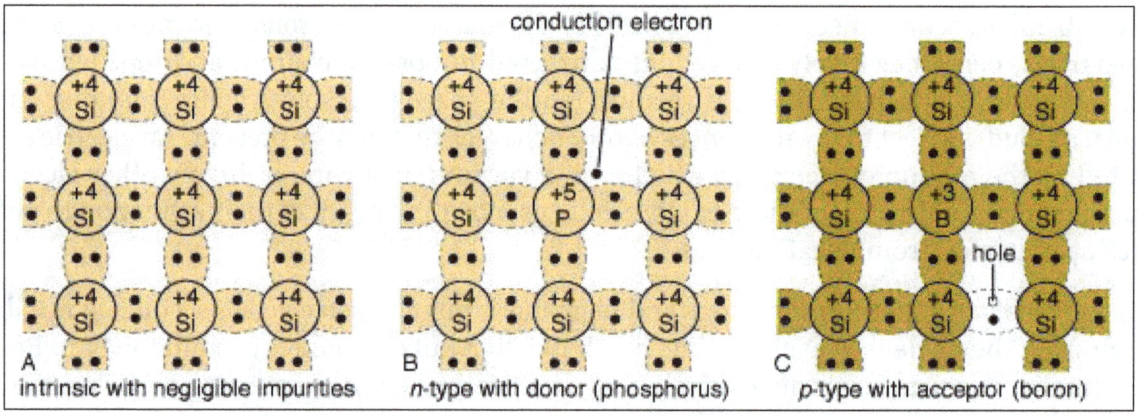

Semiconductor bonds: Three bond pictures of a semiconductor.

Semiconductors whose principal charge carriers are electrons are called n-type (n standing for negative). If the charge carriers are mainly holes, the material is p-type (p for positive). The process of substituting elements for the silicon (in this example) is called doping, while the elements are referred to as dopants. The amount of dopant that is required in practical devices is very small, ranging from about 100 dopant atoms per million silicon atoms downward to 1 per billion.

Fabrication of Semiconductors

Dopants may be added to the silicon either during the crystal growth process or later. Growth of silicon crystals begins with the preparation of extremely pure polycrystalline silicon having less than 1 dopant atom per 10 billion silicon atoms. This silicon is melted in a quartz-lined furnace. The temperature of the molten silicon is reduced to just above the melting point (1,410 °C [2,570 °F]), and a small bar (the seed) of silicon in single-crystal form is introduced into the surface of the melt. The molten silicon freezes slowly onto the seed with a crystalline structure that is continuous with the structure of the seed. The seed is slowly withdrawn, usually while rotating, under carefully controlled conditions, and it brings with it a cylindrical ingot of silicon that is a single crystal throughout. This ingot may be up to 300 mm (12 inches) in diameter and weigh up to 100 kg (220 pounds).

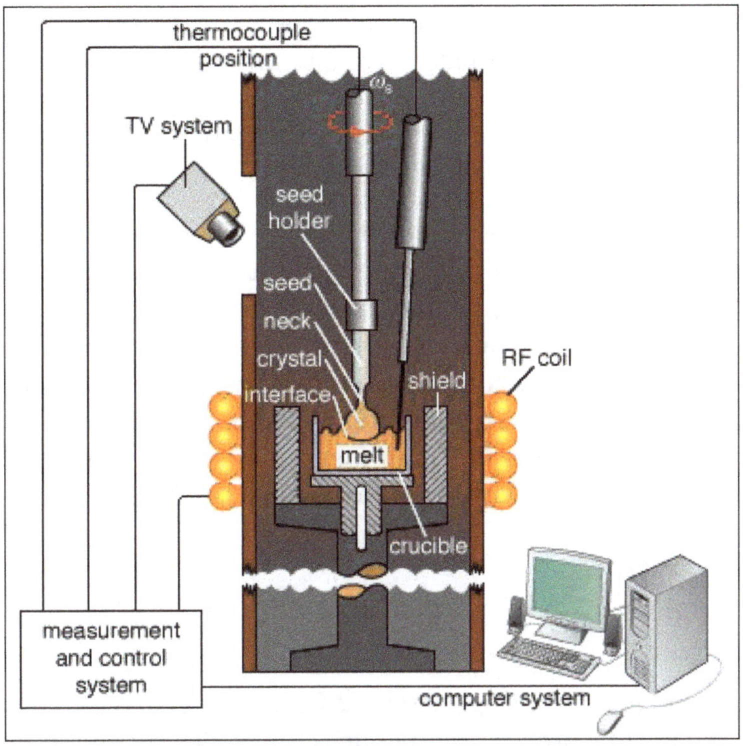

Crystal pulling using the Czochralski method: A schematic view of a modern apparatus for crystal pulling using the Czochralski method.

After growth the silicon crystal is ground to a smooth cylindrical shape and sliced into thin wafers approximately 0.6 mm (0.02 inch) thick using diamond tools. The surfaces of the wafers are polished flat by a series of successively finer abrasives until one side has a perfect mirror finish.

The process of fabricating semiconductor devices is a complex series of more than 600 sequential steps, all of which must be done with utmost precision in an environment cleaner than a hospital operating room. The objective is to add the correct dopants to the silicon in the proper amounts in the right places and to connect the transistors thus produced with thin films of metals separated by other thin films of insulating materials. The scale of lateral dimensions in integrated circuits ranged down to 0.13 μm (0.000005 inch) in 2001 and continues to decrease year by year. A high-power semiconductor device for industrial use, on the other hand, may be so large as to require a slice of silicon measuring well over 125 mm (5 inches) in diameter.

State of the Art

The importance of having a thorough, detailed understanding of all the physical effects related to materials, fabrication processes, and device structures cannot be overstated. The motion of electrons and holes in semiconductors is governed by the theory of quantum mechanics, which was developed during the 1920s and '30s as a much more comprehensive theory of the behavior of all the elementary particles that make up matter. The electrical and optical effects observed in semiconductor materials, their interactions, and the effects of temperature on them are all understood in nearly complete detail. This understanding not only makes it possible to explain quantitatively what is observed in laboratory experiments but is essential for predicting how new processes and devices work.

The research necessary to develop such a detailed theoretical and experimental body of knowledge was initiated during the late 1940s and has continued in industrial, university, and government laboratories ever since. It is now possible to design new semiconductor devices to perform in a completely predictable fashion by calculating their performance from theory and from their physical configuration, with the aid of computers.

The fabrication processes used to make real devices are not as well understood, although much has been learned. Theoretical designs incorporate the assumptions that the materials are entirely pure, that dopants exist only in the proper amounts and distributions, and that the dimensions of structures have the intended values. These assumptions are true in practice only to a limited degree. Major efforts in universities and company laboratories are focused on better understanding these issues and on developing improved computer-based modeling and process-design methods. Large sums of money are spent to provide equipment and manufacturing environments that adequately control each process step and protect the material being processed from contamination.

Basic Electronic Functions

Rectification

Rectification, or conversion of alternating current (AC) to direct current (DC). A diode, or two-terminal device, is required for this process. Semiconductor diodes consist of a crystal, part of which is n-type and part p-type. The boundary between the two parts is called a p-n junction. There is a population of holes on the p-type side of the junction and a population of electrons on the n-type side. If a negative voltage is applied to the p-type side, implying a positive voltage applied to the n-type side, the holes in the p-type region will be attracted away from the p-n junction, as will the electrons on the n-type side. A region on either side of the p-n junction will be depleted of charge carriers, thus becoming effectively an insulator. In this condition, called reverse bias, only a very small leakage current flows.

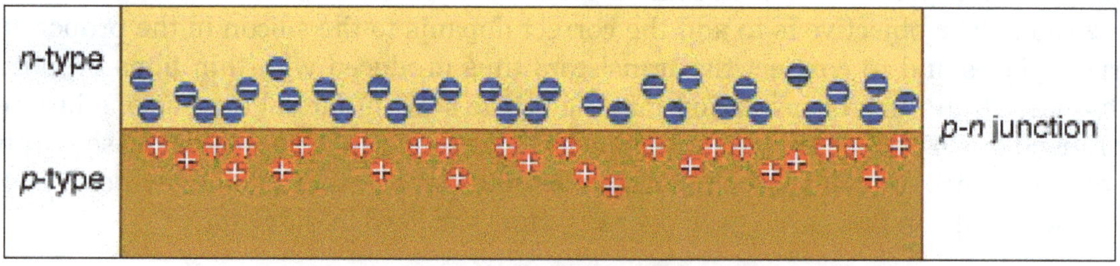

p-n junction: A barrier forms along the boundary between p-type and n-type semiconductors that is known as a p-n junction. Because electrons under ordinary conditions will flow in only one direction through such barriers, p-n junctions form the basis for creating electronic rectifiers and switches.

If these voltages are reversed, however, creating a condition called forward bias, the positive voltage on the p-type side will repel holes across the p-n junction; the negative voltage also will repel the electrons on the n-type side. Both holes and electrons will cross the p-n junction in opposing directions, creating an electric current.

Many details of the motion of holes and electrons have been omitted from this simple description, but the principle seems clear. The p-n junction in a semiconductor diode conducts current with one polarity of applied voltage but not with the other polarity. Typical small diodes will conduct about 0.1 ampere with roughly a 1.5-volt forward bias, and they will withstand 100 or more volts with negligible current flow in the reverse direction. Large industrial diodes can carry up to 5,000 amperes and can block several thousand volts.

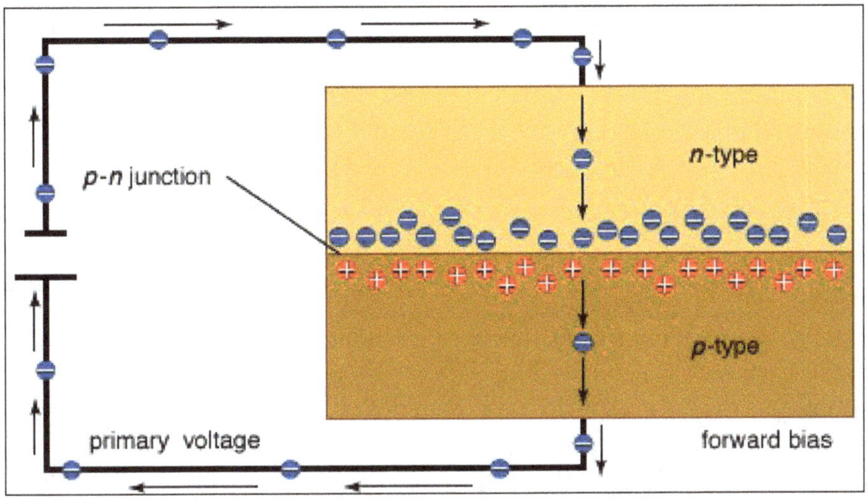

Forward-biased p-n junction: Adding a small primary voltage such that the electron source (negative terminal) is attached to the n-type semiconductor surface and the drain (positive terminal) is attached to the p-type semiconductor surface results in a small continuous current. This arrangement is referred to as being forward-biased.

Amplification

Using n-p-n Transistors

A transistor is constructed with two p-n junctions parallel and very close to one another. A typical configuration is the n-p-n transistor, which has different levels of doping in the two n-type regions and other features that improve its efficiency; in the design shown in the diagram, the n-p-n regions correspond to the source (or emitter), gate (or base), and drain (or collector) of the circuit. In normal operation, such as in an amplifier circuit, there are provisions (batteries in this case) for applying a small forward bias to the base-emitter junction and a larger reverse bias to the base-collector junction. Resistors are arranged in series with each battery to establish steady-state operating conditions, and an AC signal source is contained in the base lead. When the AC signal source is switched off, the battery in the emitter-base circuit causes a small current to flow through the series resistor and the forward-biased emitter-base junction. This results in excess electrons being present in the p-type base region of the transistor. Many more of these electrons are attracted to the collector region by the strong reverse bias on the collector than are attracted to the base connection. In an average n-p-n transistor, more than 100 electrons pass from the emitter to the collector for each 1 that passes from the emitter to the base.

When the AC signal source is switched on, the base current is increased and decreased alternately. The collector current varies in the same way but to a hundredfold larger extent; in effect, the signal has been amplified. The varying collector current through the collector series resistor causes a varying voltage drop, which may be used as the signal source for a subsequent amplifying circuit. This example employs an n-p-n transistor. With a p-n-p transistor, the action is similar except that holes are the primary charge carriers, and the voltages of the batteries and thus the direction of current are reversed.

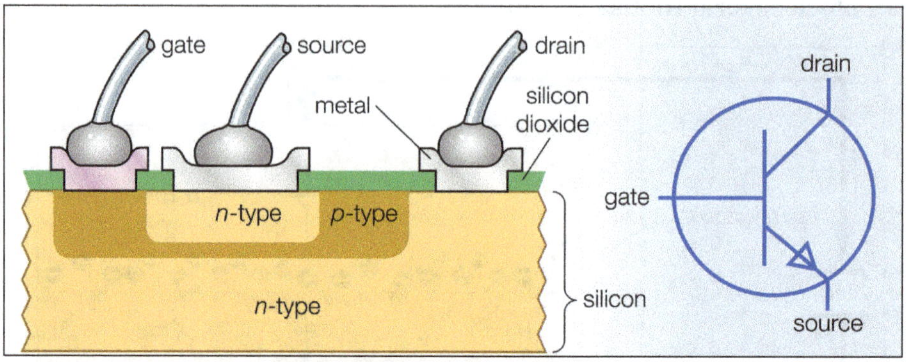

Cross section of an n-p-n transistor: An n-p-n transistor and its electronic symbol.

Using MOSFETs

Another important type of transistor developed by the early 1960s is the field-effect transistor, such as a metal-oxide-semiconductor field-effect transistor, or MOSFET. Another type, the junction field-effect transistor, works in a similar fashion but is much less frequently used. The MOSFET consists of two regions: the source (connected to the silicon substrate) and the drain of one conductivity type embedded in a body of the opposite conductivity type. The space between the source and the drain is covered by a thin layer of silicon dioxide formed by heating the silicon in an oxidizing atmosphere. A third part of the device, the gate, is a thin metal layer deposited on the silicon dioxide.

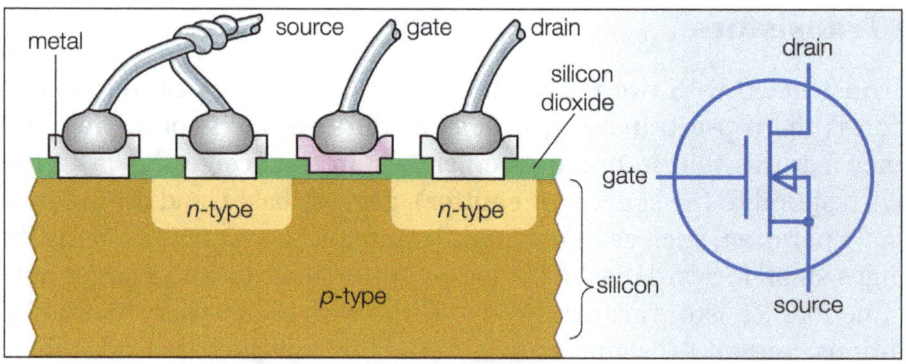

Cross-sectional view of an n-channel MOSFET and its electronic symbol.

There are several types of MOSFETs, including an n-channel type, so designated because, when it is in operation, the application of a positive voltage to the gate with respect to the p-type region causes a thin conducting region containing mostly electrons to form in the p-type region just beneath the gate. The gate voltage repels holes and attracts electrons from the p-type region, in which there are some electrons even though the principal charge carriers are holes. The thin layers of electron-rich material, the channel, connects the source and drain electrically and permits current to flow between them when the drain is biased positively with respect to the source. The amount of current is controlled by the gate voltage. Without gate voltage, no current flows, because the p-n junction around the drain region is reverse-biased and because no channel exists. MOSFETs are widely used in integrated circuits.

Coupling Amplifiers

The existence of more than one type of transistor gives the circuit designer additional freedom not available for vacuum tube circuits and allows many clever circuits to be constructed. This becomes apparent in the direct coupling of successive amplifier stages. There are many ways to couple a signal from one circuit to another. Each has its advantages and disadvantages. Consideration must be given to the voltage levels in the circuits. In cases where the voltage level at the collector of the first amplifier is different from that at the base of the second, a direct connection could not be used. A transformer could be employed for coupling, with its primary in the collector circuit of the first amplifier and it's secondary in the base circuit of the second one. However, transformers often do not exhibit uniform behavior over a wide range of frequencies, which can be a problem. Transformers also are expensive and bulky. Similarly, a capacitor could be inserted between the collector of the first amplifier and the base of the second. This works well for many applications, providing uniform coupling inexpensively over a wide frequency range. At low frequencies, capacitive coupling becomes ineffective.

The use of a p-n-p second amplifier allows direct connection between the amplifiers. If properly designed, this arrangement provides useful amplifying properties from DC to quite high frequencies. Care is required to avoid any changes in the DC operating conditions of the first amplifier; such changes will cause an amplified change in the DC conditions of the second one. Changes in temperature, in particular, can cause changes in resistor values and changes in the amplification properties of transistors. These factors must be carefully taken into account. Judicious use of feedback from later parts of a circuit to earlier ones can be utilized to stabilize such circuits or to perform various other useful functions. In negative feedback, the feedback signal is of a sense opposite to the signal present at the point in the circuit where the feedback signal is applied. While this has the effect of reducing the overall gain of the circuit, it also corrects numerous small distortions that may have occurred in the signal. For example, if the amplifier does not amplify large signals as much as small ones, the feedback from larger signals will be less, as will the reduction in gain, and the larger signals will be increased in the output of the circuit. Thus the distortion is reduced.

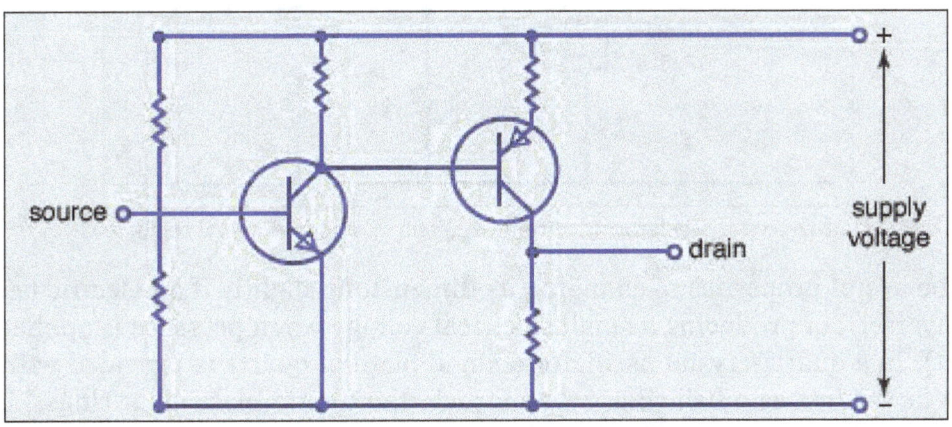

Direct coupled n-p-n and p-n-p amplifier: Electronic diagram of a direct-coupled n-p-n and p-n-p amplifier.

Oscillation

If feedback is positive, the feedback signal reinforces the original one, and an amplifier can be made to oscillate, or generate an AC signal. Such signals are needed for many purposes and are created in numerous kinds of oscillator circuits. In a tunable oscillator, such as that required for a radio receiver, the parallel combination of an inductor and a capacitor is a tuned circuit: at one frequency, and only one, the inductive effects and the capacitive effects balance. At this frequency the voltage developed across the tuned circuit is a maximum. Positive feedback is provided by the inductor in the collector circuit, which is magnetically coupled to the inductor of the tuned circuit. The connections to these inductors are arranged so that, when the collector current increases, the voltage at the base also increases, thus causing the collector current to rise further. The action of the tuned circuit reverses this sequence after a time and causes the base voltage to start to fall. This reduces the collector current; the positive feedback then further reduces the base voltage, and so on.

The circuit is in fact an amplifier whose output provides the input signal. The tuned circuit affects the feedback process in such a way that the circuit responds to an input signal at only one frequency—namely, the frequency to which the inductor and capacitor are tuned. The variable capacitor provides a way to adjust the frequency of oscillation. The output signal is obtained from the emitter resistor, through which the current rises and falls in synchrony with the collector current.

Oscillators that produce a single, accurate frequency are often needed. Such an oscillator is used in electronic watches. Other circuits in the watch count the output signals from the oscillator to determine the passage of time. These oscillators use a quartz crystal instead of a tuned circuit to establish the operating frequency.

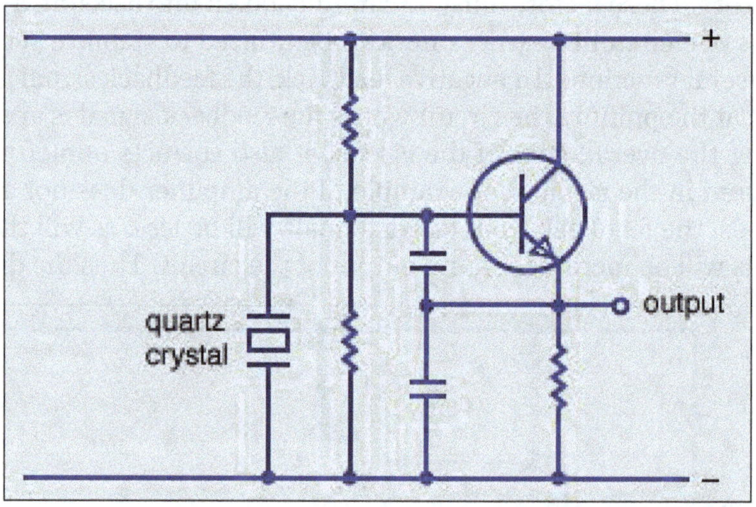

Quartz-crystal oscillator: Electronic diagram of a quartz-crystal oscillator.

Quartz has the useful properties of changing its dimensions slightly if an electric field is applied to it and, conversely, of producing a small electrical voltage when pressure is applied (the piezoelectric effect). In a quartz-crystal oscillator a small plate of quartz is provided with metal electrodes on its faces. Just as a bell rings when struck, the quartz plate also "rings," but at a very high frequency, and produces an AC voltage between the electrodes at this mechanically resonant frequency. When such a crystal is used in an oscillator, positive feedback provides energy to the quartz crystal to keep it ringing, and the oscillator output frequency is precisely controlled by the quartz crystal.

Quartz is not the only crystalline material that exhibits a piezoelectric effect, but it is used in this application because its oscillation frequency can be quite insensitive to temperature changes. Quartz-controlled oscillators are able to produce output frequencies from about 10 kilohertz to more than 200 megahertz and, in carefully controlled environments, can have a precision of one part in 100 billion, though one part in 10 million is more common.

Switching and Timing

Using Transistors

Transistors in amplifier circuits are used as linear devices; i.e., the input signal and the larger output signal are nearly exact replicas of each other. Transistors and other semiconductor devices may also be used as switches. In such applications the base or gate of a transistor, depending on the type of transistor in use, is employed as a control element to switch on or off the current between the emitter and collector or the source and drain. The purpose may be as simple as lighting an indicator lamp or it may be of a much more complex nature.

An example of a moderately sophisticated application is in a backup, or "uninterruptible," power source for a computer. Such equipment consists of a storage battery (which is normally kept charged by rectifying the power coming from the AC power line), a circuit for converting the battery power into AC, and the necessary control circuits. The control circuits monitor the voltage supplied from the power line. If this voltage varies significantly either upward or downward from its normal values, the control circuit causes the power supply lines to the computer to be switched from the incoming power line to an alternate source of AC derived from the battery.

Batteries are usually low-voltage DC sources. Consequently, their energy has to be converted to AC and applied to a transformer so as to raise the voltage to the proper level for operating the computer. The conversion from DC to AC, known as inversion, is often done with high-power transistors operated as switches. The battery is connected to the primary coil of the transformer through the transistors, first in one polarity and then in the other, at a frequency identical to the normal power-line frequency—usually 50 or 60 hertz.

The same result could in principle be obtained by operating the transistors as an oscillator powered by the battery and supplying a smoothly varying AC voltage to the transformer rather than the square pulses obtained via the switching process. This is a much less efficient procedure, however. A transistor operated as a switch is quite efficient, because in its "off" condition very little current flows at a relatively high voltage (a slight leakage through the reverse-biased collector junction), while in the "on" condition the collector-emitter voltage is very low, even though the current is large. In both conditions, the power lost is the product of the voltage and the current. Given this fact, the loss is small, because at any instant either the voltage or the current is small.

Using Thyristors

Thyristors are another important class of semiconductor devices used in switching applications. The simplest of these devices is the controlled rectifier, made of silicon. It may be regarded as two transistors connected to each other.

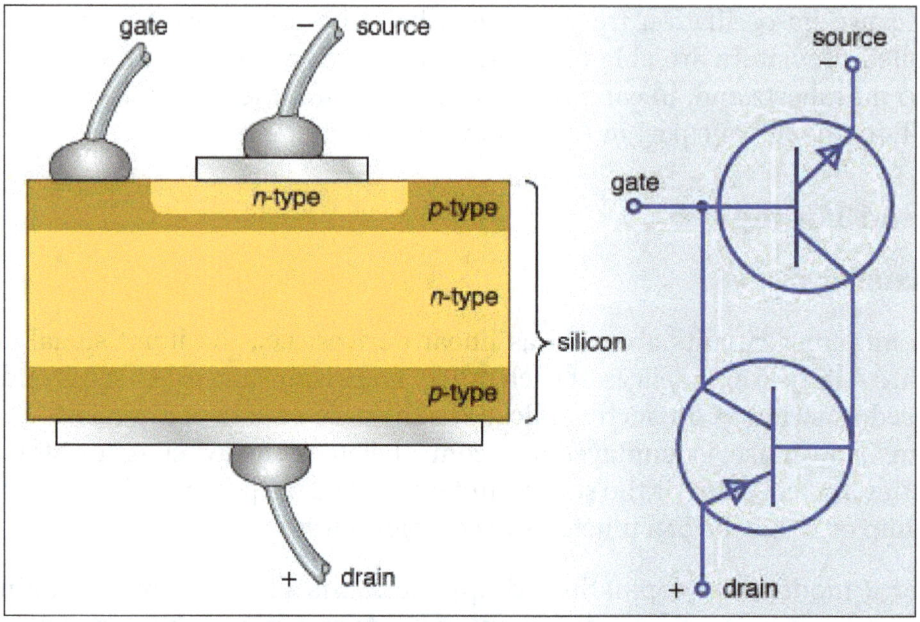

n-p-n and p-n-p thyristor: Cross section and electronic diagram for an n-p-n and p-n-p thyristor.

The device will start to conduct if a suitable amount of gate current is applied, but otherwise it will not. The gate current is the equivalent of the base current for the n-p-n transistor; the resulting larger collector current is the base current for the p-n-p transistor. The p-n-p transistor has an unusually wide base region, so its gain is small, especially at low currents. Its collector current augments the initial gate current, however. This positive feedback increases the current levels throughout the thyristor, increasing the gain of the p-n-p transistor, and at a certain point the combined currents through the n-p-n and p-n-p transistors are sufficient to maintain conduction through the device even if the gate current is removed. The transistors drive each other into a saturated condition such that the thyristor conducts a large current with a very low voltage drop, typically about one volt. The device remains in this conducting state for an arbitrary period and cannot be turned off under control of the gate. Conduction will cease if the anode polarity becomes negative with respect to the cathode.

Thyristors are thus well suited for operation in AC rather than DC circuits. They can be switched on during the appropriate half-cycle of voltage (anode positive) and will automatically switch off when the polarity reverses. A single thyristor can be used as a rectifier to produce a variable DC output from a fixed AC input. Adjustment of the DC output is made by modifying the time at which the gate current is applied after the AC voltage crosses zero and becomes the right polarity for conduction. Two thyristors connected in antiparallel (i.e., the anode of each is connected to the cathode of the other) form an AC switch, one thyristor being able to conduct on one half-cycle and the other on the alternate half-cycle. The amount of AC power delivered to the load may be adjusted to any level between zero and full power by appropriate timing of the gate signals to the two thyristors.

Thyristors are designed to handle both small and large amounts of power; the largest ones can withstand up to 5,000 volts in the "off" state and can conduct up to 2,000 amperes in the "on" state. Such a device is contained in an enclosure approximately 150 mm (6 inches) in diameter and about 30 mm (1 inch) thick fitted with external air- or water-cooling means. The power loss in the thyristor in such cases may be as much as 4 kilowatts, but the total amount of power handled may be up to 1,000 times as large. The efficiency is thus very high.

Other types of thyristors include those in which the gate is able to turn off the thyristor and those that can be switched on in either direction of current flow. The latter finds wide use in light-duty applications—for example, in variable-speed home appliances and light dimmers.

Thyristors have many applications in industrial equipment where substantial amounts of power must be controlled electronically. These applications range from transmission of electric power over long distances, which are more efficient if done as DC rather than AC, to control of heating elements in furnaces and supplying power for electronic equipment. The very large thyristors mentioned earlier are employed in power conversion for DC transmission, both from AC to DC and vice versa.

Optoelectronic Functions

Some electronic applications depend on the interactions between light and semiconductor materials mentioned in the section Optoelectronics. Such applications include the conversion of sunlight to electricity in solar cells. Most cells of this type consist of silicon diodes in specially designed enclosures to allow sunlight to illuminate them. Silicon is transparent to infrared light; this component of solar radiation passes through a solar cell without generating electricity. The waves of visible light, however, have enough energy to create hole-electron pairs (the mechanism that results in the absorption of the light). In the vicinity of the p-n junction, the holes are attracted toward the electrons on the n-type side, and the electrons are attracted to the holes on the p-type side. This constitutes a current that can be used to power small electrical appliances or to charge storage batteries.

There are special thyristors available that use light instead of a gate signal to initiate conduction. They have application in high-voltage systems wherein many thyristors in series must be employed to withstand the voltage. The practical difficulties involved in providing gate signals to all these thyristors, each at a different electrical potential, are simplified by using optical fibres (which are electrical insulators) to conduct pulses of light to the thyristors. The interaction of the light with the silicon produces carriers just as in a solar cell; these carriers provide the gate signal to switch on the thyristors.

Light-emitting diodes (LEDs) are used in many electronic systems as visual indicators. They are made from III-V compounds related to gallium arsenide; the ones that generate red light are usually composed of gallium arsenide phosphide. The central brake light on the rear of automobiles is commonly an array of red LEDs. The red light in traffic signals is also an LED application. With the availability of brilliant, low-cost blue LEDs, it is now possible to make replacements for incandescent light bulbs using a suitable mixture of coloured LEDs to provide the appropriate colour. These newer applications are driven by the need for greater reliability or electrical efficiency to justify the increase in cost.

Laser diodes, also made of III-V compounds, are used in digital audio and video disc players to read the minuscule tracks molded into the disc and containing the digitally recorded information. Lasers are employed because laser light can be focused into an extremely tiny spot of great brightness. The light scattered from the markings on the disc is detected by semiconductor photodiodes.

Electronic Engineering

Electronic engineering or electronics engineering is a form of engineering associated with electronic circuits, devices and the equipment and systems that use them. Very many people have careers in electronic engineering and find their jobs absorbing, interesting and they provide an interesting challenge.

Electronic engineering is all about creativity. The whole area of engineering is about designing, making, running, and servicing things that people need. Electronic engineering can be a particularly rewarding career: electronics engineering covers a broad spectrum of areas, all under the same banner, and this allows anyone considering a career to choose the one they want.

It is also possible to when asking the question: What is electronic Engineering, it is necessary to understand that there are many different fields and disciplines involved within the overall topic. Normally electronic engineers specialize in a particular field or area. This means that it is possible to select a particular field of interest. Some of the fields within electronic engineering include:

- Analogue electronic engineering: Analogue electronics is still a major sector within the overall electronic engineering scene. With many analogue elements still being needed, analogue circuits are still widely used. Whilst analogue engineering is not as large as it used to be many years ago before digital electronics took off in a big way, the growth in the overall electronics market has compensated for this. Analogue electronic engineering can provide a stimulating environment in which to work and a good career.
- Radio frequency engineering: Radio frequency electronic engineering has grown in its size in recent years. With many more systems using wireless links, everything from mobile phones to Wi-Fi, IoT, short range links and very much more, wireless technology is needed. Accordingly radio frequency design is in great demand and as a result it can provide a good route for a career.
- Digital development engineering: Many functions are now undertaken using digital techniques. Accordingly many digital circuits are needed and this means that some digital / logic electronic engineering is needed. That said the demand for digital engineers alone may not be as large as anticipated. The reason for this is that much logic circuitry these days is accomplished within programmable logic chips.
- Programmable logic engineering: With the complexity of many logic / digitally based circuits, an approach that is being used increasingly is one where programmable logic chips are used. FPGAs and other programmable logic chips are widely used, enabling large amounts of logic to be incorporated into programmable chips. Using high levels design languages like VHDL, etc., the design is brought to within manageable limits. Also if the design needs optimising or changing, this can be achieved by changing the logic programme. This sector of the electronic engineering arena is growing, and the tools becoming more sophisticated and very interesting to use. It can provide a very interesting and stimulating career, and t may often be linked with seem digital design for the external circuitry.

- Software engineering: There is an increasing amount of software contained within electronic products these days. As a result software engineering is becoming increasingly important. In many projects, at least two thirds of the development budget is allocated to developing the software – an indication of how this sector has grown, and hence it provides excellent opportunities for anyone wanting a career in this area.
- Systems engineering: Systems engineering is a particularly important element of the design of any item. In terms of this sector of electronic engineering, a system is any completed object. It may be a radio receiver, it may be a mobile phone, or it could be an item consisting of several individual items.

  The terms system engineering refers to the fact that this form of engineering looks at the complete object or system, comprising of smaller items, everything from boards to complete units. It looks at the operation of the overall system, ensuing that the initial requirements are correct, and that the item is finally tested to ensure that it operates to its specification, and also to the initial requirements that were placed upon it.

There are also many other, more niche areas of electronic engineering: component engineering, reliability engineering, risk management, quality assurance and many more. All are very important and need basic electronic engineering skills, but they can be applied in a different way.

Whilst most electronic engineers will tend to specialize in one area for their career, it is important to have knowledge of the other areas. This helps the engineer to interface intelligently with others from slightly different disciplines as always happens on large projects.

Take one every day example: A mobile phone base station has many elements to it, each requiring electronic engineers with different specialties. There is the radio frequency parts used for the transmitter and receiver as well as the antennas. However large amounts of software are required as the system is complicated and requires very many elements to be controlled. Signals have complex functions along with acknowledgements of messages sent and received. Software engineering is required for this. Also programmable logic is used for many of the logic requirements and this links into both software and also the logic hardware design as well as many analogue functions.

Other engineering skills including system design, installation planning, cellular coverage planners and many others will be needed as well. It can be seen that something as common as a mobile phone base station requires many engineers with a whole variety of skills.

Electronics Vs. Electrical

Electrical and electronic engineering are not just two names for the same thing! Electrical and electronic engineering may sound similar, but they are very different. Both involve moving electricity around a circuit to power useful products and machines, but that is where the similarity ends.

How do they Work?

The main difference between electrical and electronic circuits is that electrical circuits have no decision making (processing) capability, whilst electronic circuits do. An electric circuit simply powers machines with electricity. However, an electronic circuit can interpret a signal or an instruction, and perform a task to suit the circumstance. For example, a microwave oven often bleeps when it has finished cooking, to inform the user that his or her meal is ready.

Most modern appliances use a combination of electronic and electrical circuitry. A washing machine has an electrical circuit comprising a plug socket, fuse, on/off switch, heater and motor, which rotates the drum. The desired wash cycle and temperature are inputted by the user via the control panel. These instructions are interpreted by electronic circuits, which have been designed and programmed to understand what the user would like based on what buttons have been pressed. When the electronic circuit has interpreted these commands, it sends signals to the electrical circuit to operate the heater and motor, to heat and rotate the drum, for the time required.

A Difference of Scale

Most electronic components are very small, and require small direct current (DC) voltages. A single micro-processor, which will fit on the end of your finger, may contain hundreds or thousands of tiny components, some of which are only a few atoms wide! Electrical components tend to be larger, and use alternating current (AC) voltages. Whilst most electronic components operate on 3-12 volts DC, electrical appliances require 230 volts AC. In factories and power stations, however, components may require up to 11,000 volts, and most electricity travels around the UK at 400,000 volts.

Some products, such as computers, have far more electronic components than electrical components. Large industrial sites such as factories or power stations, however, have far more electrical components.

How do they Work Together?

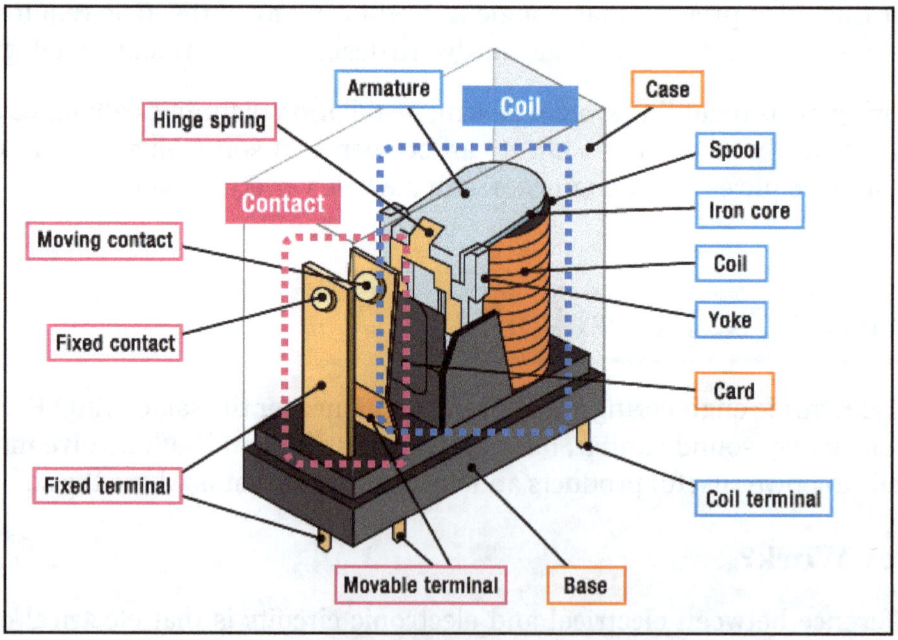

The link between electronic and electrical circuits is typically provided by relays or transistors. These are essentially switches but, rather than being pushed manually like a light switch, are operated by a small current from an electronic circuit. Therefore, a small circuit - often with many tiny components -can be used to operate much larger electrical equipment. This makes using household and industrial products safer, and means they are smaller and more energy efficient.

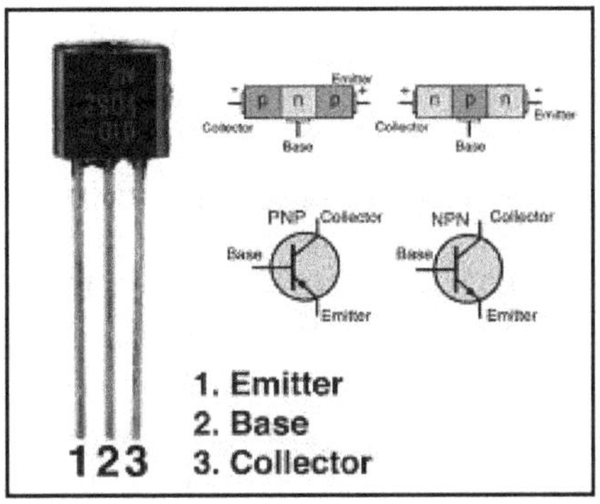

Relays are mechanical devices which, when a small current is applied from an electronic circuit, a metal contactor closes the electrical circuit, allowing a much larger current to pass. Relays were however large and unreliable, and tended to require a lot of current. After repeated use, the mov ing parts become worn and stop functioning correctly. Transistors, however, can be made much smaller and require tiny amounts of current, and have no moving parts. The transistor was invent ed in 1948, and is arguably the most important invention of the last 100 years.

2

Digital Electronics: An Overview

Digital electronics is a branch of electronics which studies circuits that operate on digital signals. Digital circuits accept and process binary data according to boolean logic. In order to completely understand digital electronics and digital circuits, it is necessary to understand the processes related to it. This chapter elucidates the varied processes and mechanisms associated with this area of study.

We know there are two types of signals, one is analog or continuous signal and the second one is Digital or discrete signal. So the science or field of research in the area of engineering is termed as Analog and Digital Electronics respectively. Now coming to the area of Digital Electronics, it is essential to understand wide range of applications from industrial electronics to the fields of communication, from micro embedded systems to military equipment. The main and perhaps the most revolutionary advantage of digital electronics are the decrease in size and the improvement in technology.

Digital Circuits

The usual advantages of digital circuits when compared to analog circuits are:

- Digital systems interface well with computers and are easy to control with software. It is often possible to add new features to a digital system without changing hardware, and to do this remotely, just by uploading new software. Design errors or bugs can be worked-around with a software upgrade, after the product is in customer hands.
- Information storage can be much easier in digital systems than in analog ones. In particular, the great noise-immunity of digital systems makes it possible to store data and retrieve it later without degradation. In an analog system, aging and wear and tear will degrade the information in storage, but in a digital system, as long as the wear and tear is below a certain level, the information can be recovered perfectly.

Robustness

One of the primary advantages of digital electronic stems from it simply being digital: robustness. The robustness stems from the fact that if the noise is less than the noise margin then the system performs as if there were no noise at all. This is a necessary and sufficient property for a circuit to be considered digital. However, if the noise exceeds this level, the circuit can give unexpected and undesired results — perhaps catastrophic incorrect results.

Digital signals can thus be regenerated to achieve lossless data transmission, within certain limits. Analog signal transmission and processing, by contrast, always introduces noise. Theoretically, there is no data-loss when copying digital data. This is a great advantage over analog systems, which faithfully reproduce every bit of noise that makes its way into the signal.

Disadvantages

Digital circuits use more energy than analog circuits to accomplish the same calculations and signal processing tasks, thus producing more heat as well. In portable or battery-powered systems this can be a major limiting factor, but in a situation where power is plentiful, a digital system is often preferred because of all the advantages listed above, especially that of (re-)programmability and ease of upgrading without requiring hardware changes.

A particular example is the cellular telephone, which being a battery-powered portable device, uses a low-power analog front-end to acquire and tune in the radio signal from the base station. The base station, being in a fixed location with access to the power grid, can afford to use a power-hungry software-defined (digital) radio technique that digitize the signal essentially at the antenna (after wideband filtering and down conversion to intermediate frequency) and performs all channelization and demodulation via software-driven calculations. Such base stations can be reprogrammed, potentially via remote control, to process the signals used in future cellular standards as those standards become available. Digital circuits are sometimes more expensive, especially in small quantities.

The world in which we live is analog, and signals from this world such as light, temperature, sound, electrical conductivity, electric and magnetic fields, and phenomena such as the flow of time, are for most practical purposes continuous and thus analog quantities rather than discrete digital ones. For a digital system to do useful things in the real world, translation from the continuous realm to the discrete digital realm must occur, resulting in quantization errors. This problem can usually be mitigated by designing the system to store enough digital data to represent the signal to the desired degree of fidelity. The Nyquist-Shannon sampling theorem provides an important guideline as to how much digital data is needed to accurately portray a given analog signal.

Fragility

Digital systems can be fragile, in that if a single piece of digital data is lost or misinterpreted, the meaning of large blocks of related data can completely change. This problem can be mitigated by designing the digital system for robustness. For example, a parity bit or other error-detecting or error-correcting code can be inserted into the signal path so that if less than a certain fraction of the data is corrupted, the system can determine that this has occurred and possibly uncorrupt the data, or ask for the corrupted data to be resent. In a state-machine, the state transition logic can be designed to catch all unused states and trigger a reset sequence or other error recovery routine. For example, it is standard practice in embedded software design to fill unused program memory with interrupt instructions that point to an error recovery routine, to help guard against a failure that corrupts the microcontroller's instruction pointer which could otherwise cause random code to be executed.

Analog Issues in Digital Circuits

Digital circuits are made from analog components, and care has to be taken in design so that the analog natures of these underlying components don't dominate over the desired digital behavior. In particular, attention must be paid to all noise and timing margins, to parasitic inductances and capacitances, to proper filtering of power and ground connections, to electromagnetic coupling amongst data lines, and many other details. Inattention to these can cause intermittent problems such as "glitches", vanishingly-fast pulses that may trigger some logic but not others, "runt pulses" that do not reach valid switching (threshold) voltages, or unexpected ("uudecoded") combinations of logic states.

A corollary of the fact that digital circuits are made from analog components is the fact that digital circuits are slower to perform calculations than analog circuits that occupy a similar amount of physical space and consume the same amount of power. However, the digital circuit will perform the calculation with much better repeatability, due to the high noise immunity of digital circuitry.

Construction

A digital circuit is often constructed from small electronic circuits called logic gates. Each logic gate represents a function of Boolean logic. A logic gate is an arrangement of electrically controlled switches. The output is an electrical flow or voltage that can, in turn, control more logic gates. Logic gates often use the fewest number of transistors in order to reduce their size, power consumption and cost, and increase their reliability. Manufactured as integrated circuits, they are the least expensive implementation when made in large volumes. They are usually designed by engineers using electronic design automation software).

Another form of digital circuit is constructed from lookup tables, (many sold as "programmable logic devices", though other kinds of PLDs exist). Lookup tables can perform all the same functions as machines based on logic gates, but lookup tables can be easily reprogrammed without changing the wiring. This means that a designer can often repair errors without changing the arrangement of wires. Therefore, in small volume products, programmable logic devices are often the preferred solution. They are usually designed by engineers using electronic design automation software.

When the volumes are medium to large, and the logic can be slow, or involves complex algorithms or sequences, often a small microcontroller is programmed to make an embedded system. These are usually programmed by software engineers.

When only one digital circuit is needed, and its design is totally customized, as for a factory production line controller, the conventional solution is a programmable logic controller, or PLC. These are usually programmed by electricians, using ladder logic.

Structure of Digital Systems

Engineers use many methods to minimize logic functions, in order to reduce the complexity, and thus the number of errors and the expense of digital circuits. The most widely used method is the application of a minimization algorithm within a CAD system, although historically, truth tables, Karnaugh Maps, and Boolean algebra have also been used.

Representations are crucial to an engineer's design of digital circuits. Some analysis methods only work with particular representations. The classical way to represent a digital circuit is with an equivalent set of logic gates. Another way, often with the least electronics, is to construct an equivalent system of electronic switches (usually transistors). One of the easiest ways is to simply have a memory containing a truth table. The inputs are fed into the address of the memory, and the data outputs of the memory become the outputs.

For automated analysis, these representations have digital file formats that can be processed by computer programs. Most digital engineers are very careful to select computer programs ("tools") with compatible file formats.

To choose representations, engineers consider types of digital systems. Most digital systems divide into "combinatorial systems" and "sequential systems". A combinatorial system always presents the same output when given the same inputs. It is basically a representation of a set of logic functions.

A sequential system is a combinatorial system with some of the outputs fed back as inputs. This makes the digital machine perform a "sequence" of operations. The simplest sequential system is probably a flip flop, a mechanism that represents a binary digit or "bit".

Sequential systems are often designed as state machines. In this way, engineers can design a system's gross behavior, and even test it in a simulation, without considering all the details of the logic functions.

Sequential systems divide into two further subcategories. "Synchronous" sequential systems change state all at once, when a "clock" signals changes state. "Asynchronous" sequential systems propagate changes whenever inputs change. Synchronous sequential systems are made of well-characterized asynchronous circuits such as flip-flops, that change only when the clock changes, and which have carefully designed timing margins.

The usual way to implement a synchronous sequential state machine is divide it into a piece of combinatorial logic and a set of flip flops called a "state register." Each time a clock signal ticks, the state register captures the feedback generated from the previous state of the combinatorial logic, and feeds it back as an unchanging input to the combinatorial part of the state machine. The fastest rate of the clock is set by the most time-consuming logic calculation in the combinatorial logic.

The state register is just a representation of a binary number. If the states in the state machine are numbered (easy to arrange), the logic function is just some logic that produces the number of the next state.

In comparison, asynchronous systems are very hard to design because all possible states, in all possible timings must be considered. The usual method is to construct a table of the minimum and maximum time that each such state can exist, and then adjust the circuit to minimize the number of such states, and force the circuit to periodically wait for all of its parts to enter a compatible state. (This is called "self-resynchronization.") Without such careful design, it is easy to accidentally produce asynchronous logic that is "unstable", that is, real electronics will have unpredictable results because of the cumulative delays caused by small variations in the values of the electronic components. Certain circuits (such as the synchronizer flip-flops, switch denouncers, and the like which allow external unsynchronized signals to enter synchronous logic circuits) are inherently asynchronous in their design and must be analyzed as such.

As of now, almost all digital machines are synchronous designs because it is much easier to create and verify a synchronous design. However, asynchronous logic is thought to be superior, if it can be made to work, because its speed is not constrained by an arbitrary clock; instead, it simply runs at the maximum speed permitted by the propagation rates of the logic gates from which it is constructed. Building an asynchronous circuit using faster parts implicitly makes the circuit "go" faster.

More generally, many digital systems are data flow machines. These are usually designed using synchronous register transfer logic, using hardware description languages such as VHDL or Verilog. In register transfer logic, binary numbers are stored in groups of flip flops called registers. The outputs of each register are a bundle of wires called a "bus" that carries that number to other calculations. A calculation is simply a piece of combinatorial logic. Each calculation also has an output bus, and these may be connected to the inputs of several registers. Sometimes a register will have a multiplexer on its input, so that it can store a number from any one of several buses. Alternatively, the outputs of several items may be connected to a bus through buffers that can turn off the output of all of the devices except one. A sequential state machine controls when each register new data from its input.

In the 1980s, some researchers discovered that almost all synchronous register-transfer machines could be converted to asynchronous designs by using first-in-first-out synchronization logic. In this scheme, the digital machine is characterized as a set of data flows. In each step of the flow, an asynchronous "synchronization circuit" determines when the outputs of that step are valid, and presents a signal that says, "grab the data" to the stages that use that stage's inputs. It turns out that just a few relatively simple synchronization circuits are needed.

The most general-purpose register-transfer logic machine is a computer. This is basically an automatic binary abacus. The control unit of a computer is usually designed as a micro program run by a micro sequencer. A micro program is much like a player-piano roll. Each table entry or "word" of the micro program commands the state of every bit that controls the computer. The sequencer then counts, and the count address the memory or combinatorial logic machine that contains the micro program. The bits from the micro program control the arithmetic logic unit, memory and other parts of the computer, including the micro sequencer itself. In this way, the complex task of designing the controls of a computer is reduced to a simpler task of programming a relatively independent collection of much simpler logic machines.

Computer architecture is a specialized engineering activity that tries to arrange the registers, calculation logic, buses and other parts of the computer in the best way for some purpose. Computer architects have applied large amounts of ingenuity to computer design to reduce the cost and increase the speed and immunity to programming errors of computers. An increasingly common goal is to reduce the power used in a battery-powered computer system, such as a cell-phone. Many computer architects serve an extended apprenticeship as micro programmers. "Specialized computers" are usually a conventional computer with a special-purpose micro program.

Automated Design Tools

To save costly engineering effort, much of the effort of designing large logic machines has been automated. The computer programs are called "electronic design automation tools" or just "EDA." Simple truth table-style descriptions of logic are often optimized with EDA that automatically produces reduced systems of logic gates or smaller lookup tables that still produce the desired outputs.

Most practical algorithms for optimizing large logic systems use algebraic manipulations or binary decision diagrams, and there are promising experiments with genetic algorithms and annealing optimizations.

To automate costly engineering effort, some EDA can take state tables that describe state machines and automatically produce a truth table for the combinatorial part of a state machine. The state table is a piece of text that lists each state, and the conditions that can exit that state.

It is common for the truth tables of such computer-generated state-machines to be optimized with logic-minimization software. This is a simple example of how complex logic machines are broken into simpler parts. Often, real logic systems are designed as a series of sub-projects, which are combined using a "tool flow". The tool flow is usually a "script", a simplified computer language that can invoke the software design tools in the right order. Tool flows for large logic systems such as microprocessors can be thousands of commands long, and combine the work of hundreds of engineers.

Writing and debugging tool flows is an established engineering specialty in companies that produce complex logic machines. The tool flow usually terminates in a detailed computer file or set of files that describe how to physically construct the logic machine. Often it consists of instructions to draw the transistors and wires on an integrated circuit or a printed circuit board.

Parts of tool flows are "debugged" by testing the outputs of simulated logic machines against expected inputs. The test tools take computer files with sets of inputs and outputs, and highlight discrepancies between the simulated behavior and the expected behavior. These test data are usually called "test vectors." Often, the test vectors are preserved and used in the factory to test that newly constructed logic machines work correctly.

Design for Testability

A large logic machine (say, with more than a hundred logical variables) can have an astronomical number of possible states. Obviously, in the factory, testing every state is impractical if testing each state takes a microsecond, and there are more states than the number of microseconds since the universe began. Unfortunately, this ridiculous-sounding case is typical.

Fortunately, large logic machines are almost always designed as assemblies of smaller logic machines. To save time, the smaller sub-machines are isolated by permanently-installed "design for test" circuitry, and are tested independently.

One common test scheme known as "scan design" moves test bits serially (one after another) from external test equipment through one or more serial shift registers known as "scan chains". Serial scans have only one or two wires to carry the data, and minimize the physical size and expense of the infrequently-used test logic.

After all the test data bits are in place, the design is reconfigured to be in "normal mode" and one or more clock pulses are applied, to test for faults (e.g. stuck-at low or stuck-at high) and capture the test result into flip-flops and/or latches in the scan shift register(s). Finally, the result of the test is shifted out to the block boundary and compared against the predicted "good machine" result.

In a board-test environment, serial to parallel testing has been formalized with a standard called

"JTAG" (named after the "Joint Testing Acting Group" that proposed it). Another common testing scheme provides a test mode that forces some part of the logic machine to enter a "test cycle." The test cycle usually exercises large independent parts of the machine.

Trade-offs

Several numbers determine the practicality of a system of digital logic. Engineers explored numerous electronic devices to get an ideal combination of speed, low cost and reliability. The cost of a logic gate is crucial. In the 1930s, the earliest digital logic systems were constructed from telephone relays because these were inexpensive and relatively reliable. After that, engineers always used the cheapest available electronic switches that could still fulfill the requirements.

The earliest integrated circuits were a happy accident. They were constructed not to save money, but to save weight, and permit the Apollo Guidance Computer to control an inertial guidance system for a spacecraft. The first integrated circuit logic gates cost nearly $50 (in 1960 dollars, when an engineer earned $10,000/year). To everyone's surprise, by the time the circuits were mass-produced, they had become the least-expensive method of constructing digital logic. Improvements in this technology have driven all subsequent improvements in cost.

With the rise of integrated circuits, reducing the absolute number of chips used represented another way to save costs. The goal of a designer is not just to make the simplest circuit, but to keep the component count down. Sometimes this results in slightly more complicated designs with respect to the underlying digital logic but nevertheless reduces the number of components, board size, and even power consumption.

For example, in some logic families, NAND gates are the simplest digital gate to build. All other logical operations can be implemented by NAND gates. If a circuit already required a single NAND gate, and a single chip normally carried four NAND gates, then the remaining gates could be used to implement other logical operations like logical and. This could eliminate the need for a separate chip containing those different types of gates.

The "reliability" of a logic gate describes its mean time between failures (MTBF). Digital machines often have millions of logic gates. Also, most digital machines are "optimized" to reduce their cost. The result is that often, the failure of a single logic gate will cause a digital machine to stop working.

Digital machines first became useful when the MTBF for a switch got above a few hundred hours. Even so, many of these machines had complex, well-rehearsed repair procedures, and would be nonfunctional for hours because a tube burned-out or a moth got stuck in a relay. Modern transistorized integrated circuit logic gates have MTBFs of nearly a trillion ($1x10^{12}$) hours, and need them because they have so many logic gates.

The "fan out" describes how many logic inputs can be controlled by a single logic output. The minimum practical fan out is about five. Modern electronic logic using CMOS transistors for switches have fan outs near fifty, and can sometimes go much higher.

The "switching speed" describes how many times per second an inverter (an electronic representation of a "logical not" function) can change from true to false and back. Faster logic can accomplish more operations in less time. Digital logic first became useful when switching speeds got above fifty hertz, because that was faster than a team of humans operating mechanical calculators. Modern electronic digital logic routinely switches at five gigahertz ($5x10^9$ hertz), and some laboratory systems switch at more than a terahertz ($1x10^{12}$ hertz).

Non-electronic Logic

It is possible to construct non-electronic digital mechanisms. In principle, any technology capable of representing discrete states and representing logic operations could be used to build mechanical logic. Hydraulic, pneumatic and mechanical versions of logic gates exist and are used in situations where electricity cannot be used. The first two types are considered under the heading of fluidics. One application of fluidic logic is in military hardware that is likely to be exposed to a nuclear electromagnetic pulse (nuclear EMP, or NEMP) that would destroy any electrical circuits.

Mechanical logic is frequently used in inexpensive controllers, such as those in washing machines. Famously, the first computer design, by Charles Babbage, was designed to use mechanical logic. Mechanical logic might also be used in very small computers that could be built by nanotechnology.

Another example is that if two particular enzymes are required to prevent the construction of a particular protein, this is the equivalent of a biological "NAND" gate. The term digital signal is used to refer to more than one concept. It can refer to discrete-time signals that are digitized or to the waveform signals in a digital system.

Discrete-time Signals

An analog signal is a datum that changes over time—say, the temperature at a given location; the depth of a certain point in a pond; or the amplitude of the voltage at some node in a circuit—that can be represented as a mathematical function, with time as the free variable (abscissa) and the signal itself as the dependent variable (ordinate). A discrete-time signal is a sampled version of an analog signal: the value of the datum is noted at fixed intervals (for example, every microsecond) rather than continuously.

If individual time values of the discrete-time signal, instead of being measured precisely (which would require an infinite number of digits), are approximated to a certain precision—which, therefore, only requires a specific number of digits—then the resultant data stream is termed a digital signal. The process of approximating the precise value within a fixed number of digits, or bits, is called quantization.

In conceptual summary, a digital signal is a quantized discrete-time signal; a discrete-time signal is a sampled analog signal. In the Digital Revolution, the usage of digital signals has increased significantly. Many modern media devices, especially the ones that connect with computers use digital signals to represent signals that were traditionally represented as continuous-time signals; cell phones, music and video players, personal video recorders, and digital cameras are examples.

In most applications, digital signals are represented as binary numbers, so their precision of quantization is measured in bits. Suppose, for example, that we wish to measure a signal to two significant decimal digits. Since seven bits, or binary digits, can record 128 discrete values (viz, from 0 to 127), those seven bits are more than sufficient to express a range of one hundred values.

Waveforms in Digital Systems

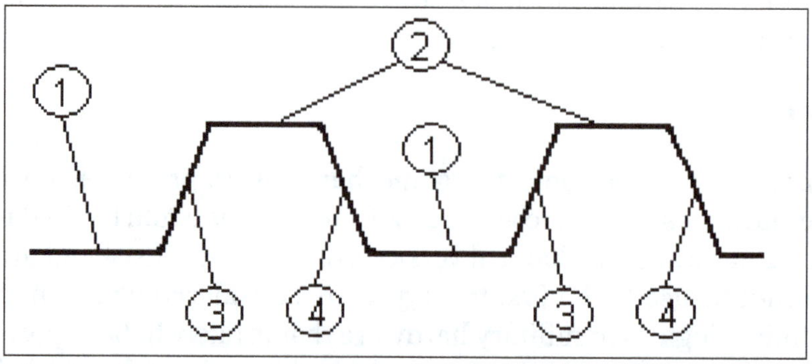

Digital signal: 1) Low level, 2) High level, 3) Rising edge, and 4) Falling edge.

In computer architecture and other digital systems, a waveform that switches between two voltage levels representing the two states of a Boolean value (0 and 1) is referred to as a digital signal, even though it is an analog voltage waveform, since it is interpreted in terms of only two levels.

The clock signal is a special digital signal that is used to synchronize digital circuits. The image shown can be considered the waveform of a clock signal. Logic changes are triggered either by the rising edge or the falling edge.

Logic Voltage Levels

The two states of a wire are usually represented by some measurement of electric current: Voltage is the most common, but current is used in some logic families. A threshold is designed for each logic family. When below that threshold, the wire is "low," when above "high". Digital circuits establish a "no man's area" or "exclusion zone" that is wider than the tolerances of the components. The circuits avoid that area, in order to avoid indeterminate results.

Hobbyist frequency counter circuit built almost entirely of TTL logic chips.

It is usual to allow some tolerance in the voltage levels used; for example, 0 to 2 volts might represent logic 0, and 3 to 5 volts logic 1. A voltage of 2 to 3 volts would be invalid and would occur only in a fault condition or during a logic level transition, as most circuits are not purely resistive, and therefore cannot instantly change voltage levels. However, few logic circuits can detect such a fault, and most will just choose to interpret the signal randomly as either a 0 or a 1. The levels represent the binary integers or logic levels of 0 and 1. In active-high logic, "low" represents binary 0 and "high" represents binary 1. Active-low logic uses the reverse representation.

Number System and Codes

Number System

A digital system can understand positional number system only where there are a few symbols called digits and these symbols represent different values depending on the position they occupy in the number. A value of each digit in a number can be determined using:

- The digit,
- The position of the digit in the number,
- The base of the number system (where base is defined as the total number of digits available in the number system).

Decimal Number System

The number system that we use in our day-to-day life is the decimal number system. Decimal number system has base 10 as it uses 10 digits from 0 to 9. In decimal number system, the successive position to the left of the decimal point represents units, tens, hundreds, thousands and so on.

Each position represents a specific power of the base 10. For example, the decimal number 1234 consists of the digit 4 in the unit's position, 3 in the tens position, 2 in the hundreds position and 1 in the thousands position and its value can be written as:

$(1\times1000) + (2\times100) + (3\times10) + (4\times1)$

$(1\times10^3) + (2\times10^2) + (3\times10^1) + (4\times10^0)$

$1000 + 200 + 30 + 1$

1234

As a computer programmer or an IT professional, you should understand the following number systems which are frequently used in computers.

S. N.	Number System and Description
1	Binary Number System. Base 2. Digits used: 0, 1.

2	Octal Number System. Base 8. Digits used: 0 to 7.
3	Hexa Decimal Number System. Base 16. Digits used: 0 to 9, Letters used: A- F.

Binary Number System

Characteristics

- Uses two digits, 0 and 1.
- Also called base 2 number system.
- Each position in a binary number represents a 0 power of the base (2). Example: 2^0.
- Last position in a binary number represents an x power of the base (2). Example: 2^x where x represents the last position - 1.

Example: Binary Number: 10101_2

Calculating Decimal Equivalent:

Step	Binary Number	Decimal Number
Step 1	10101_2	$((1 \times 2^4) + (0 \times 2^3) + (1 \times 2^2) + (0 \times 2^1) + (1 \times 2^0))_{10}$
Step 2	10101_2	$(16 + 0 + 4 + 0 + 1)_{10}$
Step 3	10101_2	21_{10}

10101_2 are normally written as 10101.

Octal Number System

Characteristics

- Uses eight digits, 0,1,2,3,4,5,6,7.
- Also called base 8 number system.
- Each position in an octal number represents a 0 power of the base (8). Example: 8^0.
- Last position in an octal number represents an x power of the base (8). Example: 8^x where x represents the last position - 1.

Example: Octal Number – 12570_8.

Calculating Decimal Equivalent:

Step	Octal Number	Decimal Number
Step 1	12570_8	$((1 \times 8^4) + (2 \times 8^3) + (5 \times 8^2) + (7 \times 8^1) + (0 \times 8^0))_{10}$

Step 2	12570_8	$(4096 + 1024 + 320 + 56 + 0)_{10}$
Step 3	12570_8	5496_{10}

12570_8 are normally written as 12570.

Hexadecimal Number System

Characteristics

- Uses 10 digits and 6 letters, 0,1,2,3,4,5,6,7,8,9,A,B,C,D,E,F.
- Letters represents numbers starting from 10. A = 10, B = 11, C = 12, D = 13, E = 14, F = 15.
- Also called base 16 number system.
- Each position in a hexadecimal number represents a 0 power of the base (16). Example 16^0.
- Last position in a hexadecimal number represents an x power of the base (16). Example 16^x where x represents the last position - 1.

Example: Hexadecimal Number: $19FDE_{16}$.

Calculating Decimal Equivalent:

Step	Hexadecimal Number	Decimal Number
Step 1	$19FDE_{16}$	$((1 \times 16^4) + (9 \times 16^3) + (F \times 16^2) + (D \times 16^1) + (E \times 16^0))_{10}$
Step 2	$19FDE_{16}$	$((1 \times 16^4) + (9 \times 16^3) + (15 \times 16^2) + (13 \times 16^1) + (14 \times 16^0))_{10}$
Step 3	$19FDE_{16}$	$(65536 + 36864 + 3840 + 208 + 14)_{10}$
Step 4	$19FDE_{16}$	106462_{10}

$19FDE_{16}$ is normally written as 19FDE.

Radix-n Number System

In radix-n number system, the numbers of symbols are restricted to value n. It can be any number other than decimal, binary, octal and hexadecimal. Table indicates Radix and number of symbols.

Table: Radix and numbers of symbols.

Radix (base)	Symbols
2 (Binary)	0,1
3	0,1,2
4	0,1,2,3
5	0,1,2,3,4
6	0,1,2,3,4,5
7	0,1,2,3,4,5,6
8 (Octal)	0,1,2,3,4,5,6,7
9	0,1,2,3,4,5,6,7,8
10 (Decimal)	0,1,2,3,4,5,6,7,8,9

......	
....	
16 (Hexadecimal)	0,1,2,3,4,5,6,7,8,9,A,B,C,D,E,F

A number represented in Radix-n system, has n symbols in the system. Let us understand counting in the Radix-n system. Table indicates counting in four popular number systems.

Decimal	Binary	Octal	Hexadecimal	Radix-5	Radix-12
0	0000	0	0	0	0
1	0001	1	1	1	1
2	0010	2	2	2	2
3	0011	3	3	3	3
4	0100	4	4	4	4
5	0101	5	5	10	5
6	0110	6	6	11	6
7	0111	7	7	12	7
8	1000	10	8	13	8
9	1001	11	9	14	9
10	1010	12	A	20	A
11	1011	13	B	21	B
12	1100	14	C	22	10
13	1101	15	D	23	11
14	1110	16	E	24	12
15	1111	17	F	30	13

To learn to count in binary, let us first consider the problem of counting a quantity of items in decimal. Let us start at 0 and count up to 9 before we run out of digits. Then start with another digit position to the left and continue counting 10 through 19, then change the symbol at left and continue counting 20 through 29 and so on. Once we reach at 99, all the two digit combinations are exhausted, so a third digit is needed in order to count from 100 through 999.

A similar situation occurs while counting in binary, except that there are only two symbols. Let us begin counting 0, 1. Let us include another digit position and continue 10, 11. With three digits, the count continues 100, 101, 110, and 111 as shown in Table. Similar counting process is repeated for Radix-n number system.

As the binary symbols 0 and 1 are exactly identical to first two decimal symbols, it is possible that confusion may result in writing of a number. For example, when a number 100 is written, are we expressing quantity four or hundred? Of course, that depends on the number system that is being used. To avoid confusion between different number systems, a number is enclosed in a parentheses and a subscript is written to indicate the radix.

e.g. $(100)_2$ = Four, $(100)_{10}$ = Hundred or $(1111)_2 = (15)_{10}$ etc.

Number Base Interconversions

Modern computers do not process decimal numbers. Instead they work with binary numbers, which use only the digit 0 and 1. Hence, we shall learn how to convert from decimal to binary, octal, hexadecimal and vice versa. People do not like working with binary numbers because they are very long when representing larger decimal quantities. Therefore, Octal, hexadecimal, numbers are used to compress long strings of binary numbers. Hence, we shall learn interconversions from one system to another as shown in figure.

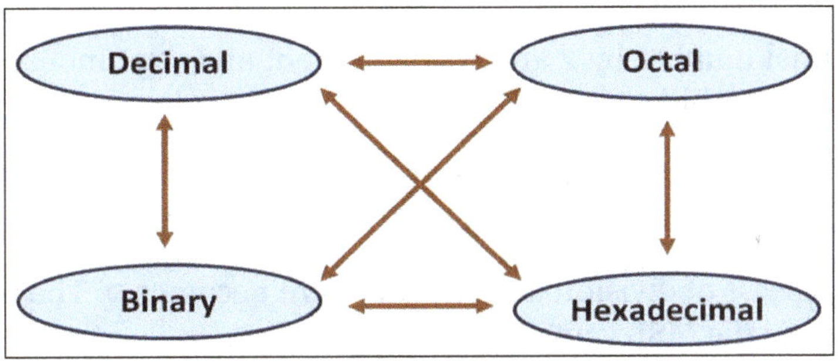

Possible number base interconversions.

Decimal to Binary Conversions (DEC to BIN)

There are two ways of converting a decimal number into a binary number. (a) Sum-of-weights method and (b) Double-dabble method.

Sum-of- weights Method

The following method can be adopted:

- Get the decimal number.
- Write the weights in exponential form or/and its value.
- Determine the set of binary weights whose sum is equal to decimal number.
- Place 1 and a 0 at appropriate positions i.e. place a 1 if weight is taken during the sum and 0 otherwise.

Example: Convert decimal number 9 into its binary equivalent.

Solution: 8 4 2 1

$(9)_{10} = 8 + 0 + 0 + 1$

$(9)_{10} = (1001)_2$

Example: Convert decimal numbers 25, 58 and 255 into their binary equivalent.

Solution: $(25)_{10} = (11001)_2$

$(58)_{10} = (111010)_2$

$(255)_{10} = (1111\ 1111)_2$

Double-dabble Method

It is a popular and most systematic way to convert decimal numbers to binary numbers. In this method a repeated division-by-2 for integers and repeated multiplication-by-2 for fractional part is used. The best way to understand the method is to follow the procedure step-by-step.

Double Dabble Procedure for Integers

- Get the decimal number.
- Divide the decimal number by 2 and write quotient and remainder. This remainder is the LSB (Least significant bit).
- Again divide the quotient obtained above by 2. This produces another quotient and remainder. The remainder is the next digit of the binary number.
- Continue this process of division until the quotient becomes 0. The remainder obtained in the final division is the MSB (most significant bit).
- Write the remainders in reverse order i.e. first remainder is LSB and last reminder is MSB.

Example: Convert decimal number 19 into its binary equivalent.

Solution: Proceed as follows:

Divisor	Dividend	Quotient	Remainder	
2	19	9	1	↑ **LSB**
2	9	4	1	
2	4	2	0	
2	2	1	0	
2	1	0	1	**MSB**

Thus $(19)_{10} = (10011)_2$.

Double –dabble Procedure for Fractional Part

- Get the fractional part.
- Multiply the decimal number by 2 and write product and carry. Note that product should be in fractions, integer part be transferred to carry. This carry is the MSB (most significant bit).
- Multiply each fractional part of product by 2 until the fractional product is zero.
- The carries generated by each multiplication form the binary number. The first carry produced is the MSB and the last one is the LSB.

Example: Convert decimal number 0.625 into its binary equivalent.

	fractional product	Carry	
0.625 ×2 =	1.25	1	MSB
0.25 ×2 =	0.50	0	↓
0.50×2 =	1.00	1	LSB

Thus, $(0.625)_{10} = (0.101)_2$.

If necessary repeated multiplication process can be continued further where greater accuracy is desired in case of nonzero product.

Binary to Decimal Conversion (BIN to DEC)

The decimal equivalent of binary number is obtained by using the fact that binary number is a weighted number. Hence we shall use once again sum- of –weights method, which is described below:

- Write the binary number.
- Write the weights of each bit under the binary number.
- If a zero appears in a bit position, delete the decimal equivalent.

Example: Convert binary number 0101 into its decimal equivalent.

0 1 0 1

Solution: $\not{8}+4+2+\not{1}=5$

$(0101)_2 = (5)_{10}$

Decimal to Octal Conversion (DEC to OCT)

The repeated division-by-8 for integer and repeated multiplication-by-8 for fractions are used for decimal to octal conversion, which is similar to the method used in conversion of decimal numbers into binary numbers (also referred as octal dabble method).

Example: Convert decimal number 127 into its octal equivalent.

Solution:

	Quotient	Remainder	
127÷8	15	7	**LSB** ↑
15÷8	1	7	
1÷8	0	1	**MSB**

Thus, $(127)_{10} = (177)_8$

$(175)_{10} = (257)_8$.

Octal to Decimal Conversion (OCT to DEC)

Conversion from octal to decimal can be done using sum of weighted numerical symbols. Following procedure may be adopted:

- Write the octal number.
- Write the weights of each symbol under that digit.
- Multiply each digit by its positional weight and then add up to obtain equivalent.

Example: Convert octal number 23 to its decimal equivalent.

Solution: $(23)_8 = 2 \times 8^1 + 3 \times 8^0$

$= 2 \times 8 + 3 \times 1$

$= 16 + 3 = (19)_{10}$

Octal to Binary Conversion (OCT to BIN)

The prime application of octal numbers is in the representation of binary number. Since it takes only one digit to represent three bits, octal numbers are much easier to 'READ' than binary numbers. Following procedure may be adopted:

- Write the octal number.
- For each octal digit write the corresponding 3-bit binary equivalent.
- The binary number may be written without a gap.

Example: Convert octal number 23 to its binary equivalent.

Solution:

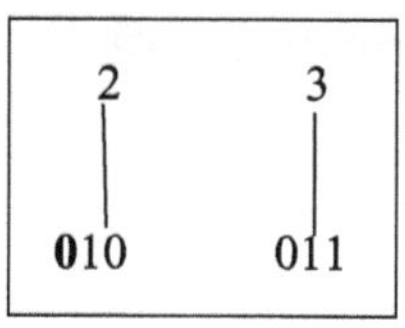

Thus $(23)_8 = (010011)_2$

Binary to Octal Conversion (BIN to OCT)

Converting a binary number to octal is a straightforward procedure, as described below:

- Beginning at a binary point, simply separate the binary number into groups of three bits towards both left as well as right.
- To complete a group, add a bit 0 if necessary.
- Replace each group of three bits with its octal equivalent.

Example: Convert binary number 11010 into its octal equivalent.

Solution: 11010 ≅ 011 010 =32

Thus $(11010)_2 = (32)_8$

Decimal to Hexadecimal Conversion (DEC to HEX)

One way to convert decimal number to Hexadecimal number is the use of Hex-dabble method. The idea is to divide integers successively by 16 and multiply fractions repetitively by 16.

Example: Convert decimal number 650 into its hexadecimal equivalent.

	Quotient	Remainder	
650 ÷ 16	40	10 (A)	LSB
40 ÷ 16	2	8	
2 ÷ 16	0	2	MSB

Thus $(650)_{10} = (28A)_{16}$

Hexadecimal to Decimal Conversion (HEX to DEC)

In hexadecimal number system each digit position corresponds to appropriate power of 16. The weights of the digit positions in hexadecimal number are as follows:

$16^3\ 16^2\ 16^1\ 16^0.\ 16^{-1}\ 16^{-2}\ 16^{-3}$

To convert Hexadecimal number to decimal number, multiply each hexadecimal digit by its weight and add the resulting product.

Example: Convert hexadecimal number 1C into its decimal equivalent.

Solution: - $1C = 1x16^1+Cx16^0$

$= 1x\ 16 + 12x1$

$= 16 + 12 = (28)_{10}$

Hexadecimal to Binary Conversion (HEX to BIN)

To convert a hexadecimal number to a binary number, first write each hexadecimal digit to its 4-bit equivalent using the binary codes, and then write the binary numbers without a gap.

Example: Convert $(9A)_{16}$ to binary equivalent.

Solution:

(9 A)

| |

$(1001\ 1010)_2$

Thus $(9A)_{16} = (1001\ 1010)_2$

Binary to Hexadecimal Conversion (BIN to HEX)

Converting a binary to hexadecimal is a straight forward procedure. Simply break the binary number into four-bit groups starting at binary point, and replace each group with the equivalent hexadecimal symbol.

Example: Convert $(101101)_2$ to Hexadecimal equivalent.

Solution: $\underset{2}{\underline{10}}\ \underset{D}{\underline{1101}}$

Thus $(10\ 1101)_2 = (2D)_{16}$

Binary Codes

There are several binary codes like weighed, un-weighted, self-complementing, cyclic etc. In this section, let us begin with the simplest BCD code.

Binary Coded Decimal (BCD) Code

The binary coded decimal (BCD) code is a weighted code. This code is found very convenient for representing digits. Each group of four bits is used to represent one decimal digit. It is also called as 8421 code. This code consists of four bits which have the weights as 8 4 2 1. The four bit combination that represents the decimal digits 0 to 9 are shown in table.

Table: The 8421 BCD code.

Decimal digit	8421 (BCD) code
0	0000
1	0001
2	0010
3	0011
4	0100
5	0101
6	0110
7	0111
8	1000
9	1001
10	0001 0000
11	0001 0001

It is realized that with four bits, sixteen different symbols (2^4) can be represented, but in 8421 BCD code only ten of these are used. The remaining six code combinations are invalid in the 8421 BCD code.

To express any decimal number in BCD code simply replaces each decimal digit by the appropriate four-bit code. Let us consider the decimal number 23,

Thus $(23)_{10} = (0010\ 0011)_{BCD}$

Advantages of BCD code over binary are that it is very simple. We do not have to perform successive division by two and write the remainder. We simply have to know the codes for the ten basic decimal digits and then we can convert any decimal number into its equivalent BCD code. The disadvantages are that it requires large number bits than the binary equivalent. BCD arithmetic implementation is difficult as compared to binary.

The BCD code has both a code and a direct binary conversion as long as the decimal numbers are integers from 0 to 9. For number greater than 9, the conversion and coding are completely different.

Weighted Code

In the weighted codes, the weights or values are assigned to the binary bits as per their bit positions. The decimal value of the code is the algebraic sum of weighted digits. In other words, the decimal number in weighted code can be written as the sum of the products of weight and the number itself. Most popular weighted codes are 8421, 5421, 7421, 2421, 5211 etc.

Table: Shows the representation of decimal number in various weighted codes.

Decimal number	Weighted codes				
	8421	5421	7421	2421	5211
0	0000	0000	0000	0000	0000
1	0001	0001	0001	0001	0001
2	0010	0010	0010	0010	0011
3	0011	0011	0011	0011	0101
4	0100	0100	0100	0100	0111
5	0101	1000	0101	1011	1000
6	0110	1001	0110	1100	1010
7	0111	1010	1000	1101	1100
8	1000	1011	1001	1110	1110
9	1001	1100	1010	1111	1111

The 8421 code utilizes the natural weights for the representation of the binary numbers hence it is popularly known as natural binary coded decimal (NBCD).

Self-complementing Code

A code is said to be self-complementing if the binary representation of a decimal number (N) in that code is 1's complement of the decimal number (9-N). Let us consider N=6. The weighted code is said to be self-complementing if N and (9-N) are 1's complement of each other. Here it is 9's complement. The 9's complement of 6 is 3.

Such codes have the property that the 9's complement of a decimal number is obtained directly by changing 1's to 0's and 0's to 1's (i.e., by complementing each bit in the pattern). The weighted codes 2421 and 5211 are self-complementing codes whereas 8421, 5421, 7421 are not self-complementing code. This could be easily verified using the Table.

Excess-3 Code

Excess-3 Code is an unweighted code, as no definite weights are assigned to bit positions. This code is sometimes used with BCD code. As the name suggests this code is obtained by adding 3 to each decimal digit and then converting it into binary. Table illustrates the excess-3 code.

Table: Excess-3 Code.

Decimal Digit	BCD Code	Excess-3 Code
0	0000	0011
1	0001	0100
2	0010	0101
3	0011	0110
4	0100	0111
5	0101	1000
6	0110	1001
7	0111	1010
8	1000	1011
9	1001	1100

For example, convert a decimal number 4 into Excess-3 code. BCD code is 0100 & excess-3 code is 0111.

The key feature of excess-3 code is that it is self-complementing. That is, the 1's complement of an excess-3 code is the excess-3 code for the 9's complement of the corresponding decimal number. The 9's complement of a decimal digit is obtained by subtracting that digit from number 9. Thus 9's complement of 3 is 6 and that of 2 is 7. The excess-3 code finds applications in arithmetic operations as it greatly simplifies the process.

Cyclic Code

While discussing cyclic codes, it is first necessary to understand the concept of Hamming distance. Hamming distance is defined as the number of places the binary bits differ in two consecutive numbers in a particular code. For example in 8421 code, hamming distance between 0 (0000) and 1 (0001) is one, as there is a change of one bit position at LSB from 0 to 1. Similarly the Hamming distance between 1 and 2 is 2. Similarly Hamming distance between 3 and 4 is 3. Thus, one can say that the Hamming distance between two successive codes of 8421 code is not constant. All cyclic codes have a unit Hamming distance property. Gray code is an example of cyclic code. It is normally used in rotary encoders, ADCs etc.

Gray Code

The Gray code is an unweighted code as no weights are assigned to bit positions. Hence, this code is not used in arithmetic operations. This code is often associated with optical encoders, a technique for converting the shaft angle into binary value. This is also referred to as cyclic code. The Gray code is basically a unit distance code, where the distance between consecutive code words is constant.

The Gray code exhibits only a one-bit change between two consecutive codes (unit Hamming distance). Table gives a comparison of binary and Gray codes. Gray code can have any number of bits like binary.

Table: Four bit Gray Code.

Decimal Digit	Binary	Gray Code
0	0000	0000
1	0001	0001
2	0010	0011
3	0011	0010
4	0100	0110
5	0101	0111
6	0110	0101
7	0111	0100
8	1000	1100
9	1001	1101
10	1010	1111
11	1011	1110
12	1100	1010
13	1101	1011
14	1110	1001
15	1111	1000

Let us understand the procedure how Gray code is generated from binary numbers:

- Write the binary number.
- The MSB in Gray code is the same as in binary number.
- Going from left to right, add each adjacent pair of binary digit to get the next Gray code digit. Discard the carries.

Let us consider the binary number 1011 to Gray code conversion:

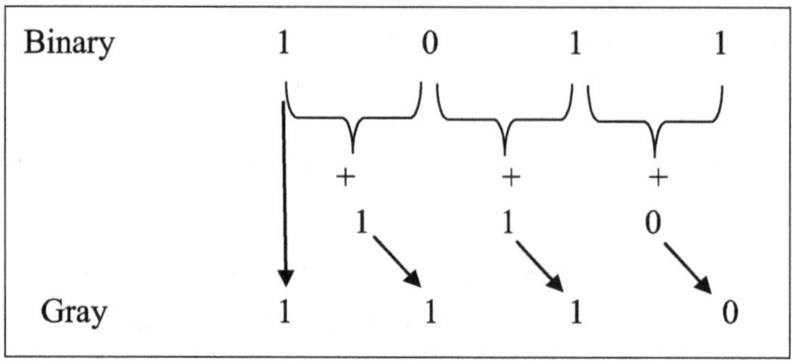

Thus, binary number 1011 is converted to gray code as 1110.

Let us understand the procedure how the Gray code is converted to a binary numbers:

- Write the Gray code.
- The MSB of binary is the same as in Gray code.
- Going from left to right, add the MSB bit of the result to the next adjacent bit of Gray code to get the next bit of binary. Discard the carries. The process is continued till the LSB is obtained.

Let us consider the binary number 1011 to Gray code conversion.

Let us convert Gray code to binary,

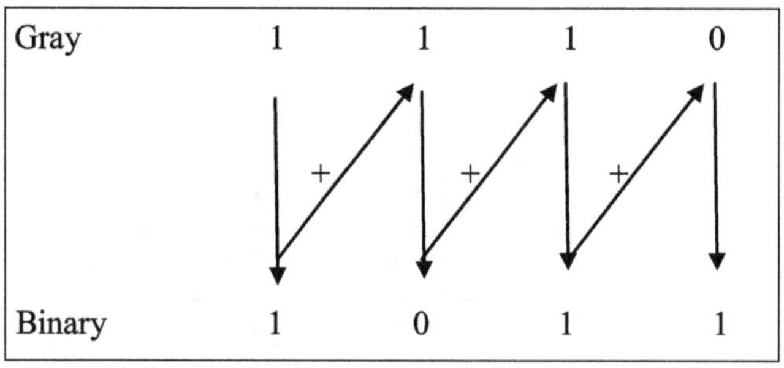

Alphanumeric Codes

All computers handle alphabets and special characters besides decimal numbers. These are normally the 26 English alphabets, the ten decimal digits and several special characters such as +,-,*, /, etc. In order to code these binary numbers, a string of binary digits is used. In modern digital computers, the number of characters has increased. Besides capital letters of the alphabet, the lower case letters are also used and several mathematical symbols such as<,>, etc. have been introduced. In such system, new coding schemes use 7 or 8 bits to code a character.

7-bit ASCII Code

The American Standard Code for Information Interchange (ASCII) code, and is extensively used for digital computers and data communication. It is a 7-bit code. It can represent 2^7 i.e. 128 different characters. It represents decimal numbers 0 to 9 by 8421 BCD code preceded by 011. The letters of the alphabet in both uppercase as well as in lowercase and other symbols, special characters and instructions for communication (or commands) are also included. The 7-bit code for each symbol has two groups of bits-3 bit group ($X_6X_5X_4$) and 4 bit group ($X_3X_2X_1X_0$).

Definitions of Control Abbreviations

ACK	Acknowledge	BEL	BELL
BS	Backspace	CAN	Cancel
CR	Carriage Return	DC_1-DC_4	Direct control
DDEL	Delete idle	DLE	Data Link Escape
EM	End of Medium	ENQ	Enquiry
EOT	End of Text	ESC	Escape
ETB	End of Transmission Block	ETX	End Text
FF	Form Feed	FS	Form Separator
GS	Group Separator	HT	Horizontal Tab
LF	Line Feed	NAK	Negative Acknowledge
NUL	Null	RS	Record Separator
SI	Shift In	SO	Shift Out
SOH	Start of Heading	STX	Start Text
SUB	Substitute	SYN	Synchronous
US	Unit Separator	VT	Vertical Tab

For example, decimal number 5 is represented by 011 0101 (35) the letter A by 100 0001 (41) and the letter m by 110 1101 (60). Note the numbers in parentheses represents hexadecimal digits.

A parity-bit can be added to the 7-bit ASCII Character Code to generate 8-bit code. A group of 8-bits is known as byte. The 8-bit ASCII code is basically an extension of 7-bit ASCII code. The ASCII code for alphabets and numerals are just the same. In addition to 128 symbols of 7-bit ASCII code, there are several symbols used for generated Tables and graphics symbols along with special mathematical symbols. These are additional 128 symbols. There are total 256 symbols in this ASCII code.

Logic Gates

Logic gates are the basic building blocks of any digital system. It is an electronic circuit having one or more than one input and only one output. The relationship between the input and the output is based on certain logic. Based on this, logic gates are named as AND gate, OR gate, NOT gate etc.

AND Gate

A circuit which performs an AND operation is shown in figure. It has n input (n >= 2) and one output:

$$Y = A\ AND\ B\ AND\ C\ ....\ N$$
$$Y = A.B.C\ ....N$$
$$Y = ABC\ ....N$$

Logic Diagram

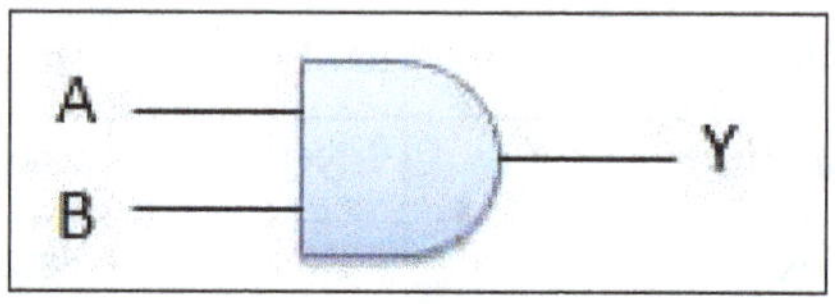

Truth Table

Inputs		outputs
A	B	AB
0	0	0
0	1	0
1	0	0
1	1	1

OR Gate

A circuit which performs an OR operation is shown in figure. It has n input (n >= 2) and one output:

$$Y = A\ OR\ B\ OR\ C\ ....\ N$$
$$Y = A + B + C\ ....N$$

Logic Diagram

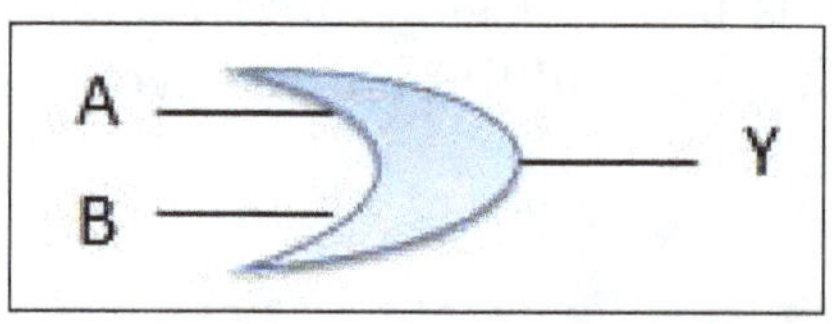

Truth Table

Inputs		Output
A	B	A+B
0	0	0
0	1	1

1	0	1
1	1	1

NOT Gate

NOT gate is also known as Inverter. It has one input A and one output Y:

$$Y = NOT\ A$$

$$Y = \bar{A}$$

Logic Diagram

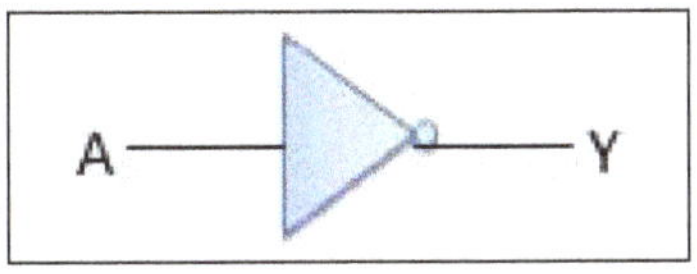

Truth Table

Inputs	Output
A	B
0	1
1	0

NAND Gate

A NOT-AND operation is known as NAND operation. It has n input (n >= 2) and one output:

$$Y = A\ NOT\ AND\ B\ NOT\ AND\ C\ ....\ N$$

$$Y = A\ NAND\ B\ NAND\ C\ ....N$$

Logic Diagram

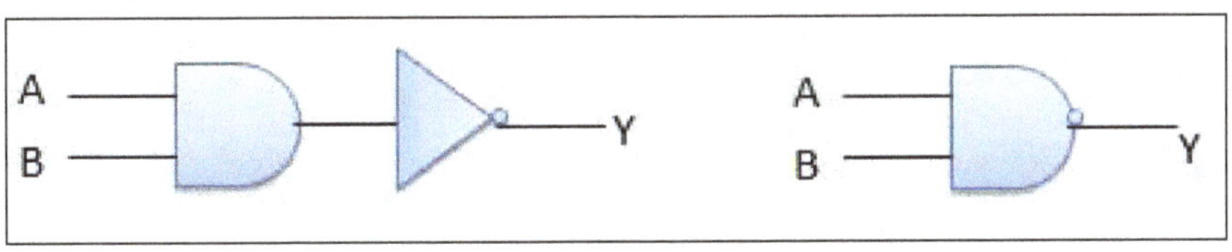

Truth Table

Inputs		output
A	B	$\overline{AB}$
0	0	1
0	1	1

1	0	1
1	1	0

NOR Gate

A NOT-OR operation is known as NOR operation. It has n input (n >= 2) and one output.

$$Y = A\ NOT\ OR\ B\ NOT\ OR\ C........\ N$$

$$Y = A\ NOR\ B\ NOR\ C\ ....\ N$$

Logic Diagram

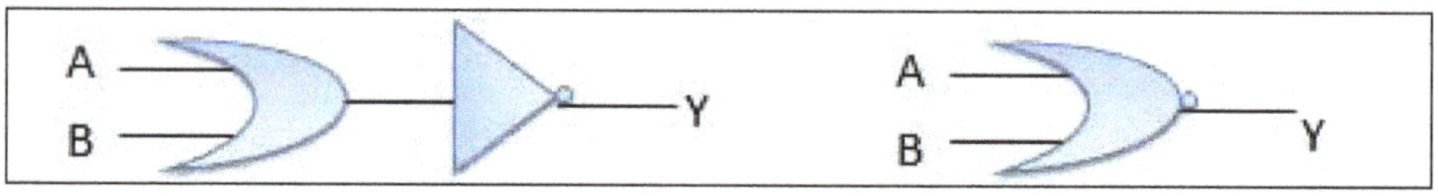

Truth Table

Inputs		Output
A	B	$\overline{A+B}$
0	0	1
0	1	0
1	0	0
1	1	0

XOR Gate

XOR or Ex-OR gate is a special type of gate. It can be used in the half adder, full adder and subtractor. The exclusive-OR gate is abbreviated as EX-OR gate or sometime as X-OR gate. It has n input (n >= 2) and one output:

$$Y = A\ XOR\ B\ XOR\ C\ ....\ N$$

$$Y = A \oplus B \oplus C\ ....N$$

$$Y = \overline{AB} + \overline{AB}$$

Logic Diagram

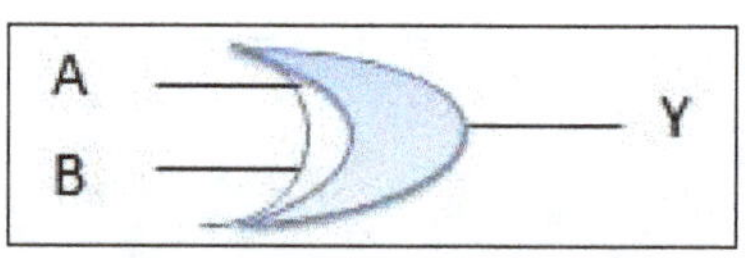

Truth Table

Inputs		Output
A	B	$A \oplus B$

0	0	0
0	1	1
1	0	1
1	1	0

XNOR Gate

XNOR gate is a special type of gate. It can be used in the half adder, full adder and subtractor. The exclusive-NOR gate is abbreviated as EX-NOR gate or sometime as X-NOR gate. It has n input (n >= 2) and one output:

$$Y = A \ XOR \ B \ XOR \ C \ .... \ N$$

$$Y = A \ominus B \ominus C \ ....N$$

$$Y = \overline{A}\,\overline{B + AB}$$

Logic Diagram

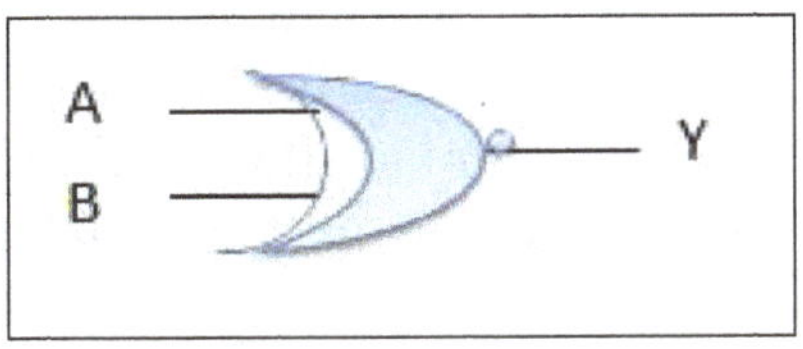

Truth Table

Inputs		Output
A	B	$A \ominus B$
0	0	1
0	1	0
1	0	0
1	1	1

Special Logic Gates

Gates with Open Collector

In case of TTL logic family, some NAND require an external resistor, called the pull-up resistor. Pull up resistor us connected between the output and the DC power supply to get desired logic function. In case of MOS family open drain gates are used. These logic gates have capability of providing an ANDing operation when outputs without using an AND gate. This connection is is popularly known as wire-AND system. Figure indicates Wired AND gate using open collector NAND gates. The output of wire-AND system is,

Y = AB'. CD'.EF'

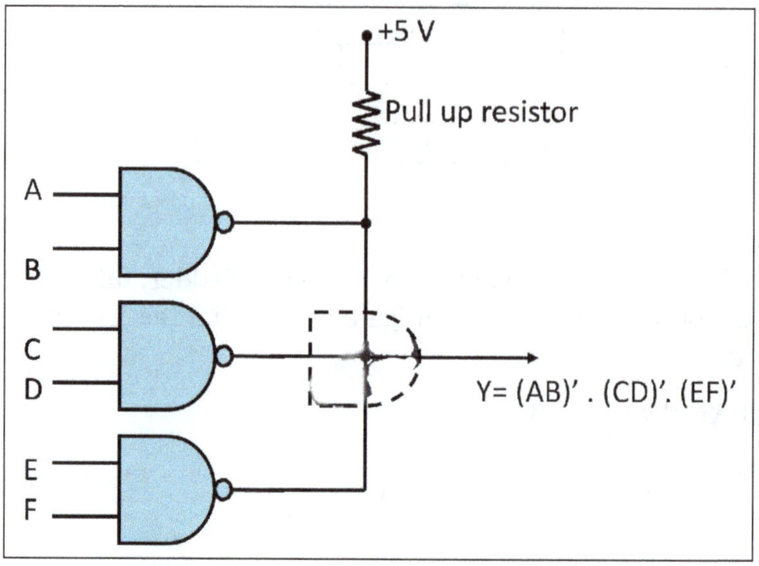

Wire-AND system using open collector NAND devices.

The open collector outputs of several NAND gates can be connected together to form then Wired-AND function. In this example, three open collector NAND gate outputs are wired together as shown in this figure. One single pull up resistor is sufficient for this interconnection. Figure shows similar circuit configuration for Wire-ANDing using open collector NOT gates.

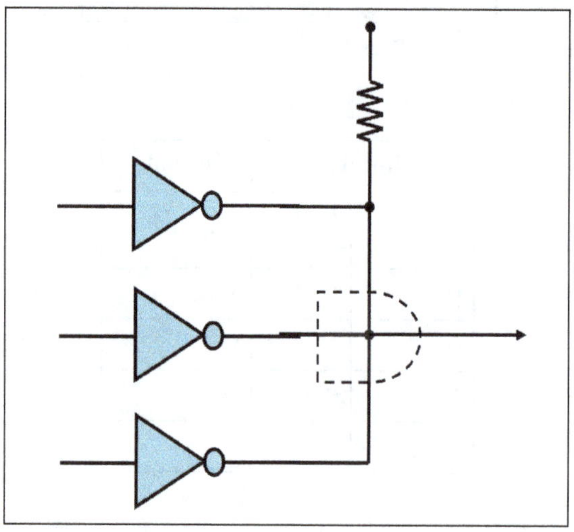

Wire-AND system using open collector NOT devices.

The output Y of wire- AND ing of all the inputs is,

Y= (A)'. (B)'. (C)'

Tristate Gates

In digital electronics, tri-state (three states) logic allows three possible output states namely the logic '1' state, the logic '0' state and a high-impedance state. This logic gate allows sharing same output line or lines. The high-impedance state is controlled by an external ENABLE input. The ENABLE input decides whether the tri-state gate is active (0 or 1) or in the high-impedance state.

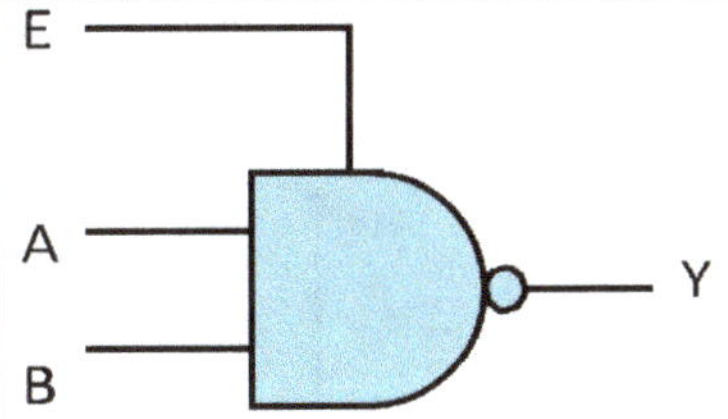

E	A	B	Y
0	0	0	Z
0	0	1	Z
0	1	0	Z
0	1	1	Z
1	0	0	1
1	0	1	1
1	1	0	1
1	1	1	0

Tri-state gate with active HIGH enable.

Figure above show when Enable input is HIGH, the NAND function provides 0 or 1 depending upon the input combination. If enable is LOW, the NAND gate output will provide high-impedance state independent of the input combinations. Let us now consider NAND gate with the active LOW enable input.

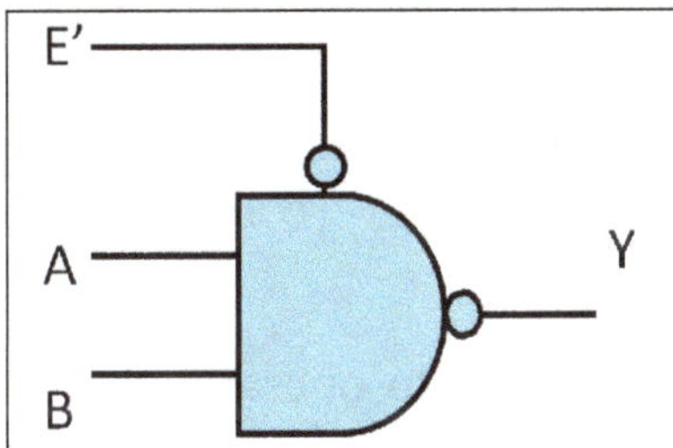

E	A	B	Y
0	0	0	1
0	0	1	1
0	1	0	1
0	1	1	0
1	0	0	Z
1	0	1	Z
1	1	0	Z
1	1	1	Z

Tri-state gate with active LOW enable.

The high-impedance state is controlled by an external ENABLE input. The ENABLE input decides whether the gate is active or in the high-impedance state. When active, it can be '0' or '1' depending upon input conditions. In this gate when Enable input is LOW, the NAND function provides 0 or 1 depending upon the input combination. If enable is HIGH, the NAND gate output will provide high-impedance state independent of the input combinations. Let us consider the interconnections of tristate inverters together to form a common bus as shown in figure.

All NOT gates are having single input and single output with enable input. The outputs are connected together to form a common bus. Each NOT gate has a separate active HIGH Enable input. This arrangement allows only one of them to be enabled at a time. HIGH ENABLE INPUT activates the respective NOT gate with a condition that only one of them is enabled at a time.

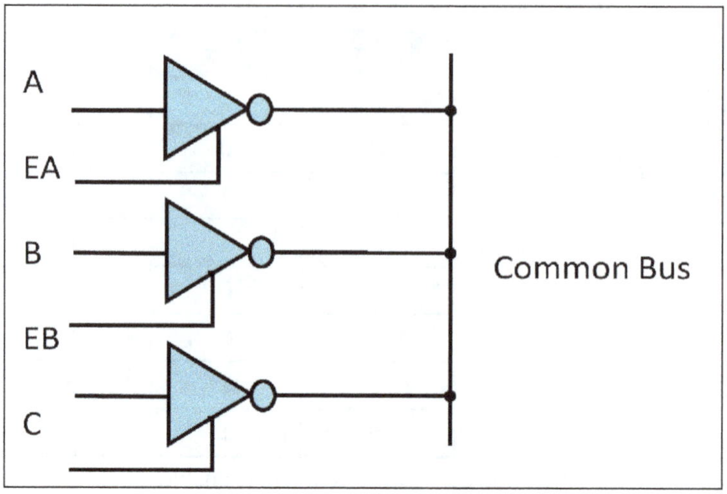

Tristate inverter.

Schmitt Gates

A standard NOT gate or NAND gate is usually made up from transistor as a switching device. Transistor cannot switch from one state to another instantly; there is some delay in switching between the states. Transistor is basically a current amplifier and can also operate in linear region. Any small changes at its input level may cause a change in its output level and can even switch ON or OFF several times if there is any noise present in the system. One way to overcome this problem is to use Schmitt inverter (NOT or NAND gates).

Schmitt gate is basically a comparator with hysteresis. It is designed to operate when its input signal goes above an "Upper Threshold Point (UTP)", the output changes and goes to LOW level and remain in the same state until the input signal falls below the "Lower Threshold Point (LTP)". Once again the output changes and goes HIGH. In short, a Schmitt gate has built in Hysteresis to overcome the problem of noise.

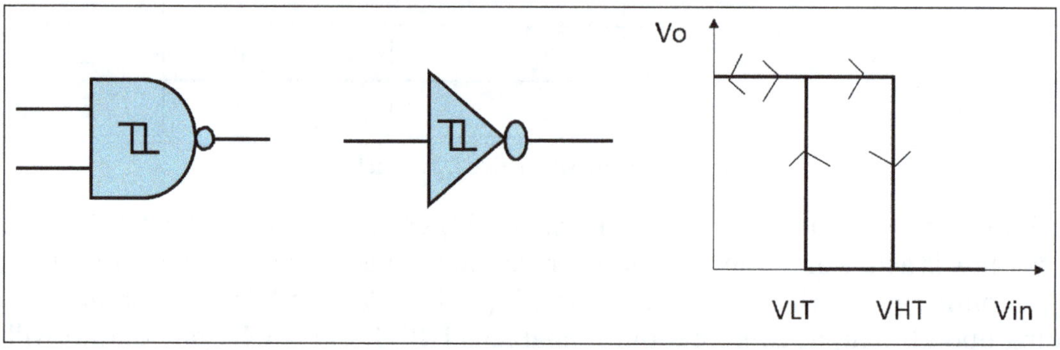

Figure shows NAND and NOT gate with built-in hysteresis. These gates are referred to as Schmitt gates. This gate interprets slowly changing input voltages according to two threshold voltages. Lower & upper threshold voltages are normally used for LOW to HIGH and HIGH to LOW transitions respectively. Schmitt gates are distinguished from conventional gates by the small 'hysteresis' symbol of the B –H loop for a ferromagnetic material. Hysteresis increases noise immunity of the digital system.

This clean ON/OFF switching makes the Schmitt gate ideal for switching any slow-rising or slow falling input signal. An important application of Schmitt gate is in oscillators i.e. sine to square wave converter or a clock.

Buffers

A buffer has only one input and one output. Normal logic gates have a limited load-driving capability. A logic gate can be treated as signal amplifiers, regardless of what logic function they may perform. For this purpose, a special logic gate called a buffer is designed. A buffer has a larger load-driving capability than a normal logic gate. It could be an inverting or non-inverting buffer with a single input, a single output and an Enable input. Figure shows the symbols and functional tables of inverting and no inverting buffers of the tri-state type. Buffer is also popularly known as "Driver". Buffers are usually tri-state devices preferred for designing bus-oriented system.

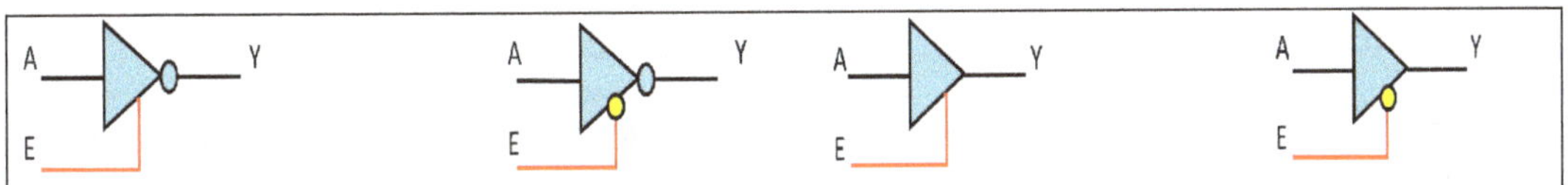

E	A	Y
0	X	Z
1	0	1
1	1	0

E'	A	Y
1	X	Z
0	0	1
0	1	0

E	A	Y
0	X	Z
1	0	0
1	1	1

E'	A	Y
1	X	Z
0	0	0
0	1	1

Inverting and non-inverting Buffers.

A tristate buffer is a useful device that allows us to control when current passes through the device and when it does not. A tri-state buffer has two inputs: a data input a and a enable input E. When the enable input is active i.e. at logic 1 then the output is same as input. If enable input is 0 then the output is in tri-state. In many applications we need tri-state buffers with active LOW enable input.

Transceivers

It is kind of buffer gate capable of both transmit and receive. A transceiver is a combination of transmitter and receiver in a single IC package. It is capable of driving bidirectional systems. A transceiver is a bidirectional buffer capable of driving bidirectional system with additional direction control/ENABLE inputs. It allows flow of data in both directions, depending upon the status of the control inputs. Transceivers can be popularly used to construct bus-oriented systems. Figure indicates the circuit symbols of inverting and non-inverting transceivers.

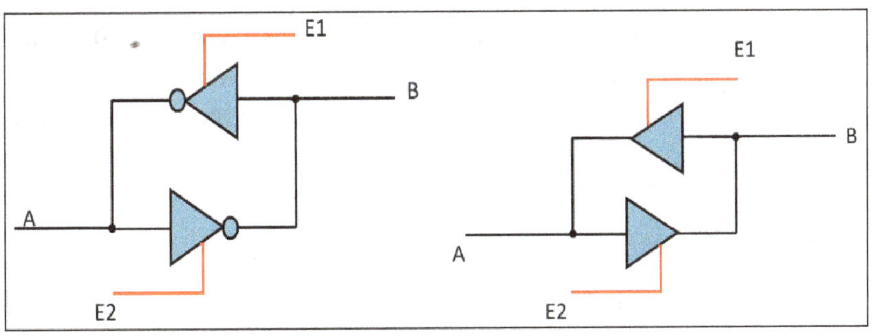

Inverting and non-inverting transceiver.

In case of inverting transceiver, when active HIGH input is applied at Enable input E1, data is transferred from A to B in complemented form and when Enable input E2=1, then the data is transferred from B to A in complemented form. Where as in non-inverting transceiver, data is transmitted or received in un-complemented form.

AOI Gate

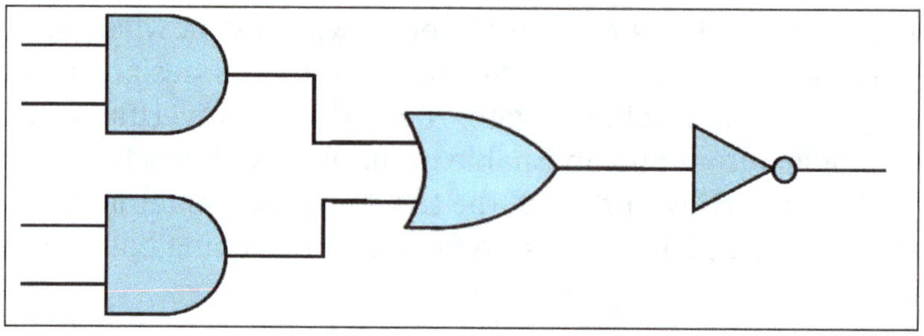

Figure indicates the special arrangement of three basic logic gates, which is popularly known as AOI (AND-OR-Invert) gate. It is used in many applications of combinational systems.

Pulsed Operation of Logic Gates

In many applications, the inputs to a gate are not stationary levels but changing dynamically between two voltage levels frequently. These types of changing voltage or logic levels give rise to Pulse waveform. Let us now study the response of different logic gates with pulse input waveform.

Pulsed Operation of NOT Gate

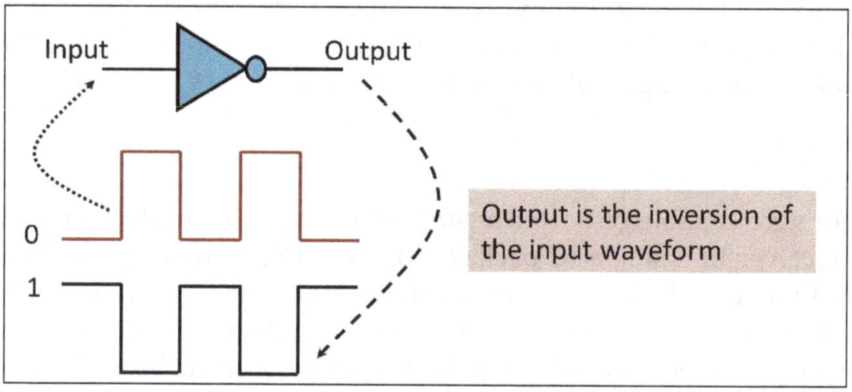

Response of NOT gate to Pulse input.

Let us now study the response of NOT gate to the pulse input waveform. It is important to note that the all logic gates obey the truth table irrespective of the static or dynamic inputs. For a NOT gate if the input is at logic 0 then output will be in logic 1 and vice a versa. The output is inverted as compared to input.

Pulsed Operation of AND Gate

In order to study the pulse response, it is important to look at the input of the gate with respect to each other to determine the output level at any time. Let us consider two AND gates with one of the input is fixed and tied to either 0 or 1 and pulse waveform is applied to the other input as shown in figure. When one of the inputs is 1, the pulse appears as it at output and output is 0 when one of the inputs is 0.

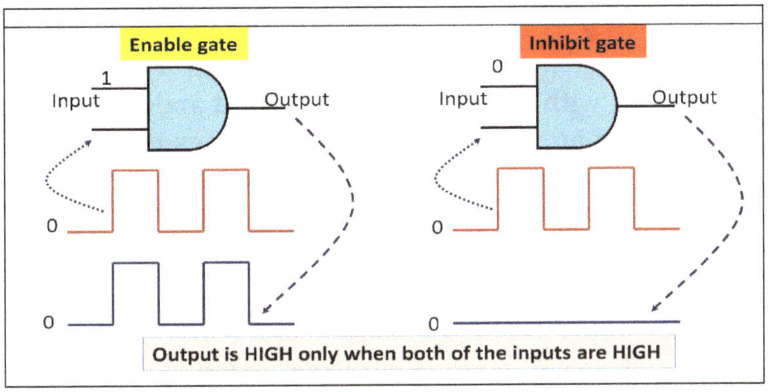

Response of AND gate to Pulse input.

Pulsed Operation of OR Gate

Let us consider two OR gates with one of the input is fixed and tied to either 0 or 1 and pulse wave-form is applied to the other input as shown in figure. When one of the inputs is 0, the pulse appears as it at output and output is 1 when one of the inputs is 1.

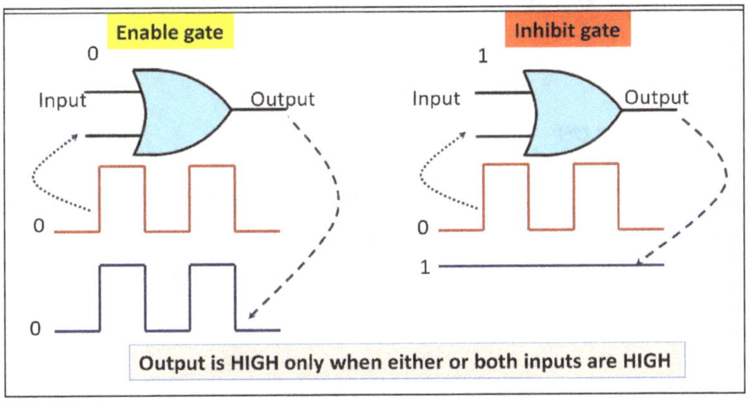

Response of OR gate to Pulse input.

Pulsed Operation of NAND Gate

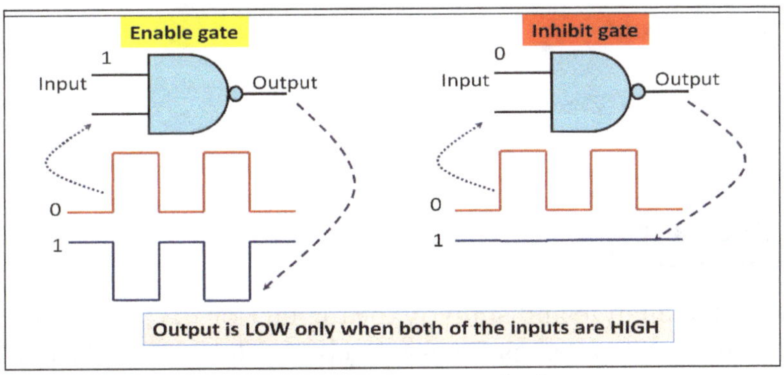

Response of NAND gate to Pulse input.

Let us consider two NAND gates with one of the input is fixed and tied to either 0 or 1 and pulse waveform is applied to the other input as shown in figure. When one of the inputs is 1, the pulse appears inverted at output and output is 1 when one of the inputs is 0.

Pulsed Operation of NOR Gate

Let us consider two NOR gates with one of the input is fixed and tied to either 0 or 1 and pulse waveform is applied to the other input as shown in figure. When one of the inputs is 0, the pulse inverted at output and output is 0 when one of the inputs is 1.

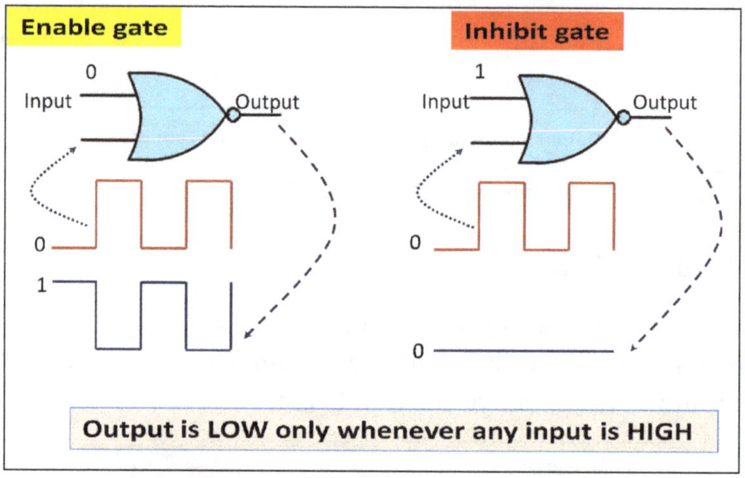

Response of NOR gate to Pulse input.

Applications of Logic Gates

The applications of logic gates are countless. Logic gates are used in many household and industrial applications. Every digital gadget including PC, mobile phones, tablets, calculators, digital watches etc. Let us now list very simple applications of Logic gates. One must choose the right logic gate for the application:

- Safety lock,
- Freezer Warning alarm,
- Enable gate for counter,
- Industrial plant control,
- Burglar Alarm etc.

Boolean Algebra

Boolean Algebra is used to analyze and simplify the digital (logic) circuits. It uses only the binary numbers i.e. 0 and 1. It is also called as Binary Algebra or logical Algebra. Boolean algebra was invented by George Boole in 1854.

Rule in Boolean Algebra

Following are the important rules used in Boolean algebra:

- Variable used can have only two values. Binary 1 for HIGH and Binary 0 for LOW.
- Complement of a variable is represented by an over bar (-). Thus, complement of variable B is represented as $\overline{B}$. Thus if B = 0 then $\overline{B}$ = 1 and B = 1 then $\overline{B}$ = 0.
- ORing of the variables is represented by a plus (+) sign between them. For example ORing of A, B, C is represented as A + B + C.
- Logical ANDing of the two or more variable is represented by writing a dot between them such as A.B.C. Sometime the dot may be omitted like ABC.

Boolean Laws

There are six types of Boolean Laws:

Commutative Law

Any binary operation which satisfies the following expression is referred to as commutative operation:

$$(1.)A.B = B.A \quad (2.)A + B = B + A$$

Commutative law states that changing the sequence of the variables does not have any effect on the output of a logic circuit.

Associative Law

This law states that the order in which the logic operations are performed is irrelevant as their effect is the same:

$$(1.)(A.B).C = A.(B.C) \quad (2.)(A + B) + C = A + (B + C)$$

Distributive Law

Distributive law states the following condition:

$$A.(B + C) = A.B + A.C$$

AND Law

These laws use the AND operation. Therefore they are called as AND laws:

$$(1.)A.0 = 0 \qquad (2.)A.1 = A$$

$$(3.)A.A = A \qquad (4.)A.\overline{A} = 0$$

OR Law

These laws use the OR operation. Therefore they are called as OR laws:

$$(1.) A + 0 = A \qquad (2.) A + 1 = 1$$

$$(3.) A + A = A \qquad (4.) A + \overline{A} = 1$$

Inversion Law

This law uses the NOT operation. The inversion law states that double inversion of a variable result in the original variable itself:

$$\overline{\overline{A}} = A$$

As well as the logic symbols "0" and "1" being used to represent a digital input or output, we can also use them as constants for a permanently "Open" or "Closed" circuit or contact respectively. A set of rules or Laws of Boolean Algebra expressions have been invented to help reduce the number of logic gates needed to perform a particular logic operation resulting in a list of functions or theorems known commonly as the Laws of Boolean Algebra.

Boolean Algebra is the mathematics we use to analyze digital gates and circuits. We can use these "Laws of Boolean" to both reduce and simplify a complex Boolean expression in an attempt to reduce the number of logic gates required. Boolean Algebra is therefore a system of mathematics based on logic that has its own set of rules or laws which are used to define and reduce Boolean expressions.

The variables used in Boolean Algebra only have one of two possible values, a logic "0" and a logic "1" but an expression can have an infinite number of variables all labelled individually to represent inputs to the expression, For example, variables A, B, C etc., giving us a logical expression of A + B = C, but each variable can ONLY be a 0 or a 1. Examples of these individual laws of Boolean, rules and theorems for Boolean Algebra are given in the following table.

Truth Tables for the Laws of Boolean

Boolean Expression	Description	Equivalent Switching Circuit	Boolean Algebra Law or Rule
A + 1 = 1	A in parallel with closed = "CLOSED"	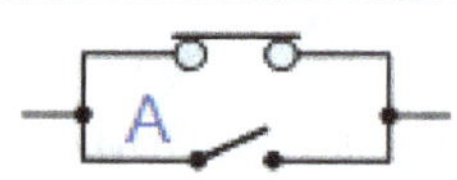	Annulment
A + 0 = A	A in parallel with open = "A"		Identity
A. 1 = A	A in series with closed = "A"	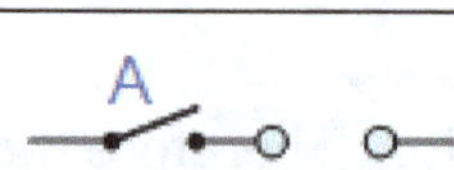	Identity
A. 0 = 0	A in series with open = "OPEN"	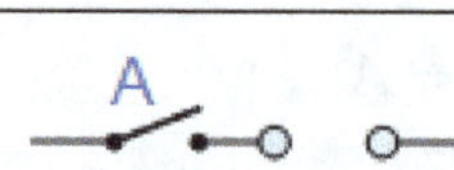	Annulment

A + A = A	A in parallel with A = "A"	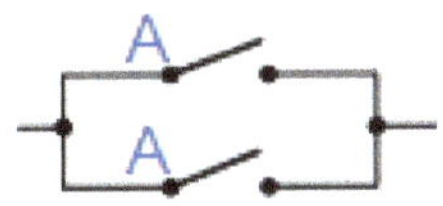	Idempotent
A. A = A	A in series with A = "A"	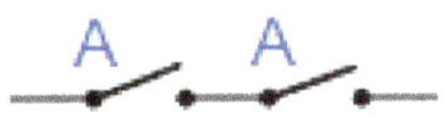	Idempotent
NOT $\overline{A}$ = A	NOT NOT A (double negative) = "A"		Double Negation
A + $\overline{A}$ = 1	A in parallel with NOT A = "CLOSED"	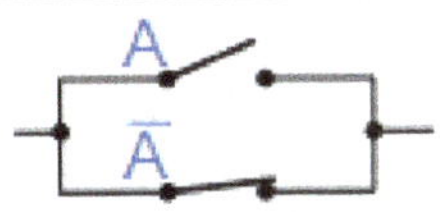	Complement
A. $\overline{A}$ = 0	A in series with NOT A = "OPEN"		Complement
A+B = B+A	A in parallel with B = B in parallel with A	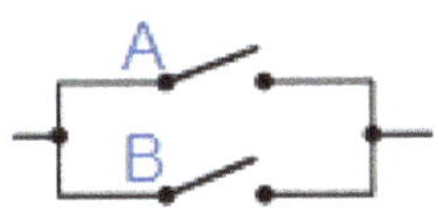	Commutative
A.B = B.A	A in series with B = B in series with A		Commutative
A+B = A.B	Invert and replace OR with AND		de Morgan's Theorem
A.B = A+B	Invert and replace AND with OR		de Morgan's Theorem

The basic Laws of Boolean Algebra that relate to the Commutative Law allowing a change in position for addition and multiplication, the Associative Law allowing the removal of brackets for addition and multiplication, as well as the Distributive Law allowing the factoring of an expression, are the same as in ordinary algebra.

Each of the Boolean Laws above are given with just a single or two variables, but the number of variables defined by a single law is not limited to this as there can be an infinite number of variables as inputs too the expression. These Boolean laws detailed above can be used to prove any given Boolean expression as well as for simplifying complicated digital circuits.

- Annulment Law: A term AND´ed with a "0" equals 0 or OR´ed with a "1" will equal 1.

  A. 0 = 0 A variable AND'ed with 0 is always equal to 0.

  A + 1 = 1 A variable OR'ed with 1 is always equal to 1.

- Identity Law: A term OR´ed with a "0" or AND´ed with a "1" will always equal that term.

  A + 0 = A A variable OR'ed with 0 is always equal to the variable.

  A. 1 = A A variable AND'ed with 1 is always equal to the variable.

- Idempotent Law: An input that is AND´ed or OR´ed with itself is equal to that input.

  A + A = A A variable OR'ed with itself is always equal to the variable.

A. A = A A variable AND'ed with itself is always equal to the variable.

- Complement Law: A term AND´ed with its complement equals "0" and a term OR´ed with its complement equals "1".

  $A.\overline{A} = 0$ A variable AND'ed with its complement is always equal to 0.

  $A + \overline{A} = 1$ A variable OR'ed with its complement is always equal to 1.

- Commutative Law: The order of application of two separate terms is not important.

  A. B = B. A The order in which two variables are AND'ed makes no difference.

  A + B = B + A The order in which two variables are OR'ed makes no difference.

- Double Negation Law: A term that is inverted twice is equal to the original term.

  $\overline{\overline{A}} = A$ A double complement of a variable is always equal to the variable.

de Morgan´s Theorem

There are two "de Morgan´s" rules or theorems:

- Two separate terms NOR´ed together is the same as the two terms inverted (Complement) and AND´ed for example: $\overline{A+B} = \overline{A}.\overline{B}$.
- Two separate terms NAND´ed together is the same as the two terms inverted (Complement) and OR´ed for example: $\overline{A.B} = \overline{A}+\overline{B}$.

Other algebraic Laws of Boolean not Detailed above Include:

- Boolean Postulates: While not Boolean Laws in their own right, these are a set of Mathematical Laws which can be used in the simplification of Boolean Expressions.

  0. 0 = 0 A 0 AND'ed with itself is always equal to 0.

  1. 1 = 1 A 1 AND'ed with itself is always equal to 1.

  1. 0 = 0 A 1 AND'ed with a 0 is equal to 0.

  0 + 0 = 0 A 0 OR'ed with itself is always equal to 0.

  1 + 1 = 1 A 1 OR'ed with itself is always equal to 1.

  1 + 0 = 1 A 1 OR'ed with a 0 is equal to 1.

  $\overline{1} = 0$ The Inverse (Complement) of a 1 is always equal to 0.

  $\overline{0} = 1$ The Inverse (Complement) of a 0 is always equal to 1.

- Distributive Law: This law permits the multiplying or factoring out of an expression.

  A (B + C) = A.B + A.C (OR Distributive Law).

A + (B.C) = (A + B). (A + C) (AND Distributive Law).

- Absorptive Law: This law enables a reduction in a complicated expression to a simpler one by absorbing like terms.

  A + (A.B) = (A.1) + (A.B) = A (1 + B) = A (OR Absorption Law).

  A (A + B) = (A + 0). (A + B) = A + (0.B) = A (AND Absorption Law).

- Associative Law: This law allows the removal of brackets from an expression and regrouping of the variables.

  A + (B + C) = (A + B) + C = A + B + C (OR Associate Law).

  A (B.C) = (A.B) C = A. B. C (AND Associate Law).

Boolean Algebra Functions

Using the information above, simple 2-input AND, OR and NOT Gates can be represented by 16 possible functions as shown in the following table.

Function	Description	Expression
1.	NULL	0
2.	IDENTITY	1
3.	Input A	A
4.	Input B	B
5.	NOT A	$\overline{A}$
6.	NOT B	$\overline{B}$
7.	A AND B (AND)	A. B
8.	A AND NOT B	$A.\overline{B}$
9.	NOT A AND B	$\overline{A}.B$
10.	NOT AND (NAND)	$\overline{A.B}$
11.	A OR B (OR)	A + B
12.	A OR NOT B	$A+\overline{B}$
13.	NOT A OR B	$\overline{A}+B$
14.	NOT OR (NOR)	$\overline{A.B}$
15.	Exclusive-OR	$A.\overline{B}+\overline{A}.B$
16.	Exclusive-NOR	$A.B+\overline{A}.\overline{B}$

Laws of Boolean Algebra

Using the above laws, simplify the following expression: (A + B) (A + C).

Q = (A + B). (A + C)	
A.A + A.C + A.B + B.C	– Distributive law
A + A.C + A.B + B.C	– Idempotent AND law (A.A = A)
A (1 + C) + A.B + B.C	– Distributive law
A.1 + A.B + B.C	– Identity OR law (1 + C = 1)
A (1 + B) + B.C	– Distributive law
A.1 + B.C	– Identity OR law (1 + B = 1)
Q =A + (B.C)	– Identity AND law (A.1 = A)

Then the expression: (A + B) (A + C) can be simplified to A + (B.C) as in the Distributive law.

Encoder

Numbers are usually coded in one form or another so as to represent or use it as required. In modern communication system, data communication involves a process encoding and decoding. Encoding is the process of converting human readable characters into binary information. Decoding is the reverse process of conversion of binary or encoded format into human readable code. This process is also referred to as code conversion.

There are numerous encoding and decoding methods exist. The oldest code of all, originally employed in the landline communication for telegraph application was the Morse code. The code used by almost all computers is the ASCII (American Standard Code for Information Interchange) code. ASCII utilizes uppercase, lowercase alphabetic characters, numeral, punctuation marks and common symbols. In addition, digital system designers use Binary, BCD, Octal Hexadecimal, Decimal, Gray, Excess-3 code and several other codes. In data communication, a special form of encoding is used to convert binary digits too high to low or low to high transitions.

Encoding and decoding are also used in context with the process of analog to digital conversion or digital to analog conversion. In this sense, code conversion term can apply to any form of data including text, images, audio, video, multimedia, computer codes, and sensor signals. Encoding should not be confused with encryption.

Code Conversion

Modern digital system like calculators, mobile phones, fax machines, computers and other digital gadgets use binary numbers 1 and 0 for processing and storage. However, our language of communication is in the form of decimal numbers and alphabetic characters. It is difficult for human being to understand information represented in binary. Therefore, there is a need of a group of devices to convert human readable language to machine or binary language. To achieve this task, several binary codes are developed. The process of generating binary codes is known as encoding and the reverse process is known as decoding. Let us consider a simple example to demonstrate the role of encoder and decoder. Figure indicates simple block diagram of generalized digital system.

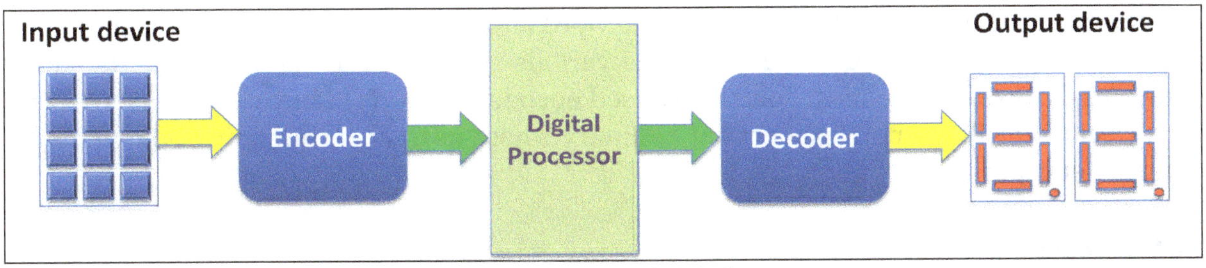

Simple block diagram of Digital system.

A digital system cannot be complete without encoders and decoders. It contains five basic building blocks. Keyboard and seven segment display are the input and output devices respectively. The encoders essentially receive the input from the keyboard. Many keyboards use encoders to produce a machine recognizable code i.e. binary and decoder accepts the processed binary code and converts back in the format useful for output devices like seven segment display, CRT, Printers etc.

BCD, binary, Octal, Hexadecimal, ASCII and decimal are the most common codes used in the digital systems. Thus encoders and decoders play an important role in code conversion process in digital system. One can make comparison of the above system with the functional blocks of hand held calculator.

Basics of Encoder

The process of converting from human readable code to machine readable code i.e. binary is known as encoding. An encoder is a combinational circuit that coverts more familiar numbers, symbols or character into binary code. An encoder has a number of input lines but only one of them is activated at a time representing a digit or character and produces a binary code depending on which input is activated. Figure is the logic symbol of encoder with 'm' inputs and 'n' outputs. In short, it is multiple inputs and multiple outputs device with proper conversion system. Note that encoder performs the reverse operation of the decoder.

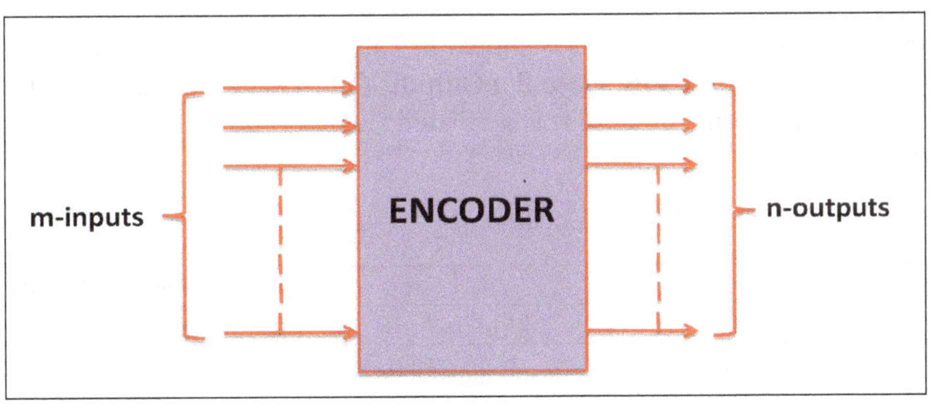

Logic Symbol of Encoder.

An encoder has 'm' number of input lines and 'n' number of output lines. The numbers of outputs (n) are always less than number of inputs (m). Here the number of inputs can have following relation,

$$m \leq 2^n$$

Thus, the encoder accepts an m-bit input digital word and converts it into an n-bit another digital word. Encoders can also be devised to encode various symbols and characters for the keypads and keyboards. Some of the most commonly used encoders are – (1) Linear encoders are octal to binary, Decimal to BCD and Hexadecimal to binary where normal encoding is implemented and (2) Priority encoders.

Liner Encoders

Linear / normal encoders accepts m inputs, only one is activated at a given time and produces nbit output code. As the numbers of output lines are 'n', it is possible to encode maximum up to 2n inputs. In many digital systems, octal, decimal or hexadecimal numbers are used as inputs.

Octal to Binary Encoder

An octal to binary encoder is also known as 8-line to 3-line encoder. It accepts 8- inputs and produces a 3-bit output corresponding to the activated octal input. Figure shows the logic symbol of Octal to binary encoder.

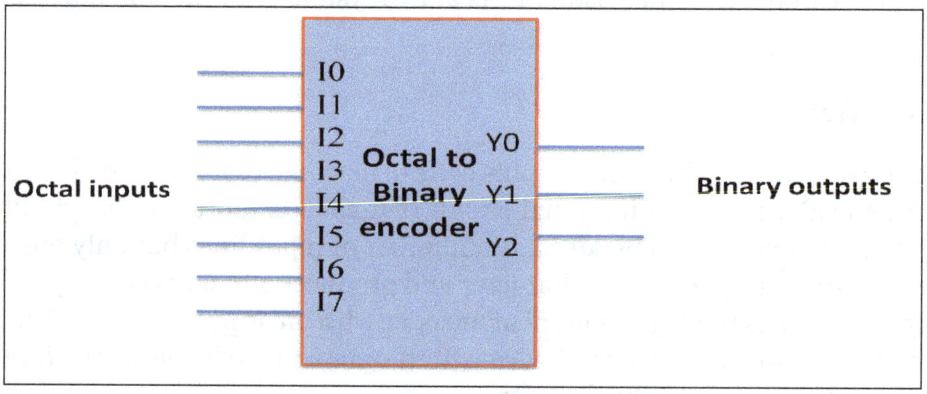

Logic symbol of Octal to binary encoder.

The truth table of octal to binary encoder is shown in Table. There are eight inputs corresponding 8 octal inputs (I0, I1, I2, I3, I4, I5, I6, I7) and 3 Outputs (Y0, Y1, Y2). Let us look at the table carefully. Note that in encoders only one input is activated (logic 1) and other inputs are not activated (i.e. at logic 0).

TABLE: Truth – Table for Octal Binary Encoder.

Octal Inputs								Binary Outputs		
I0	I1	I2	I3	I4	I5	I6	I7	Y2	Y1	Y0
1	0	0	0	0	0	0	0	0	0	0
0	1	0	0	0	0	0	0	0	0	1

0	0	1	0	0	0	0	0	0	1	0
0	0	0	1	0	0	0	0	0	1	1
0	0	0	0	1	0	0	0	0	0	0
0	0	0	0	0	1	0	0	1	0	1
0	0	0	0	0	0	1	0	1	1	0
0	0	0	0	0	0	0	1	1	1	1

In the truth table, the input variable I0 represents the least significant digit (LSD) and and I7 represents most significant digit (MSD). Similarly, in the outputs Y0 represents least significant bit (LSB) and Y2 is the most significant bit (MSB). The truth table includes only all valid combinations of the inputs. The valid combinations are those which have exactly one input equal to logic 1 while all other inputs are logic 0's.

As the number of inputs are 8, K-maps cannot be used to derive the Boolean expressions. The Boolean expression can be directly derived from the truth table by visual inspection. Let us obtain the Boolean expression for each output.

Output Y0 is 1 if any of the inputs I1 or I3 or I5 or I7 is 1. Then, the Boolean expression:

Y0 = I1 +I3 +I5 +I7

Similarly,

Y1 = I2 +I3 +I6 +I7

And,

Y2 = I4 +I5 +I6 +I7

From these Boolean expressions, the Octal to Binary Encoder can be implemented by using simply three 4 input Or gates. Figure indicates the logic diagram for Octal to Binary Encoder.

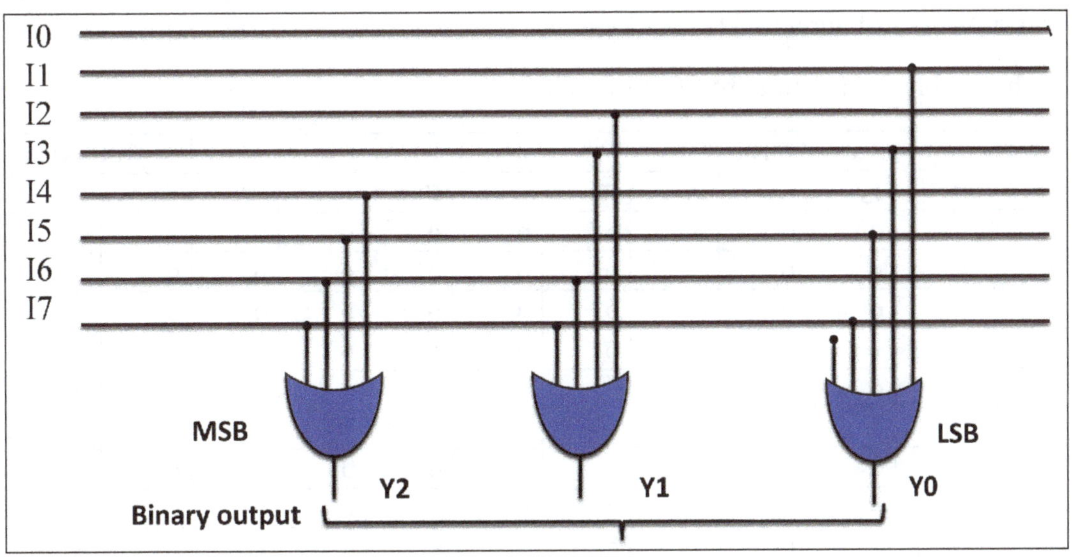

This octal to binary encoder design has few drawbacks. In the encoder, at a time only one input must be active. (1) If number of active inputs is greater than one then the output is incorrect. (2) Similarly, when no inputs are active then all outputs are low which is also similar to the output generated by active input I0. To resolve first problem, one must design and use priority encoder where the output corresponding to input with higher weightage is generated. The solution to the second drawback is to generate additional output signal to check the validity of inputs. If no inputs are active, data is invalid.

Decimal to BCD Encoder

A decimal to BCD (binary coded decimal) encoder is also known as 10-line to 4-line encoder. It accepts 10- inputs and produces a4-bit output corresponding to the activated decimal input. Figure shows the logic symbol of decimal to BCD encoder.

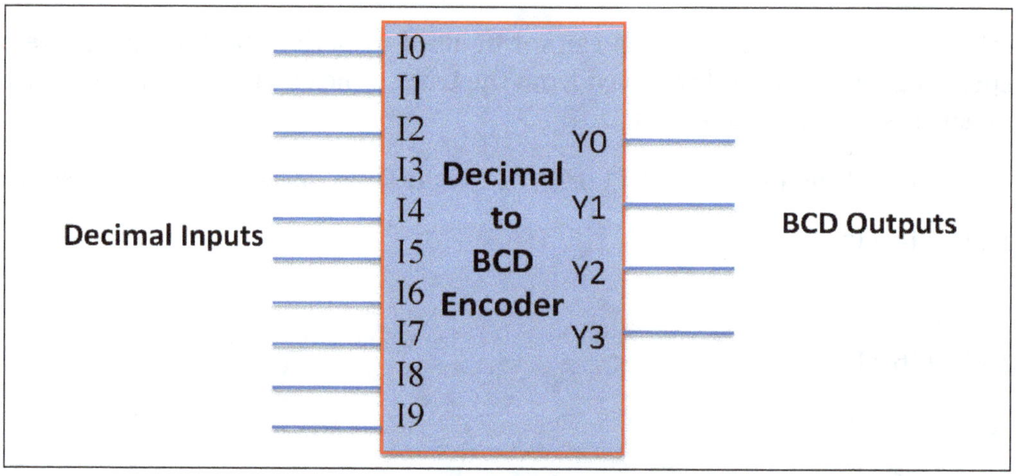

The truth table of Decimal to BCD encoder is shown in table. There are ten inputs corresponding 10 decimal inputs (I0, I1, I2, I3, I4, I5, I6, I7, I8, I9) and 4 Outputs (Y0, Y1, Y2, Y3). Let us look at the table carefully. Note the encoder assumption that only one of the inputs is activated (logic 1) and other inputs are not activated (i.e. at logic 0).

Table: Truth – Table for Decimal to BCD Encoder.

Decimal Inputs											BCD Outputs		
I0	I1	I2	I3	I4	I5	I6	I7	I8	I9	Y3	Y2	Y1	Y0
1	0	0	0	0	0	0	0	0	0	0	0	0	0
0	1	0	0	0	0	0	0	0	0	0	0	0	1
0	0	1	0	0	0	0	0	0	0	0	0	1	0
0	0	0	1	0	0	0	0	0	0	0	0	1	1
0	0	0	0	1	0	0	0	0	0	0	0	0	0
0	0	0	0	0	1	0	0	0	0	0	0	0	1
0	0	0	0	0	0	1	0	0	0	0	0	1	0
0	0	0	0	0	0	0	1	0	0	0	0	1	1
0	0	0	0	0	0	0	0	1	0	1	1	0	0
0	0	0	0	0	0	0	0	0	1	1	1	0	1

In the truth table, the input variable Io represents the least significant digit (LSD) and and I9 represents most significant digit (MSD). Similarly, in the outputs Yo represents least significant bit (LSB) and Y3 is the most significant bit (MSB). The truth table includes only all valid combinations of the inputs. The valid combinations are those which have exactly one input equal to logic 1 while all other inputs are logic o's.

As the number of inputs is 10, K-maps cannot be used to derive the Boolean expressions. The Boolean expression can be directly derived from the truth table by visual inspection. Let us obtain the Boolean expression for each output.

Output Yo is 1 if any of the inputs I1 or I3 or I5 or I7 is 1. Then, the Boolean expression,

$$Y0 = I1 + I3 + I5 + I7 + I9$$

Similarly,

$$Y1 = I2 + I3 + I6 + I7$$

$$Y2 = I4 + I5 + I6 + I7$$

$$Y3 = I8 + I9$$

From these Boolean expressions, the Decimal to BCD Encoder can be implemented by using simply three 4 OR gates. Figure indicates the logic diagram for Decimal to BCD Encoder.

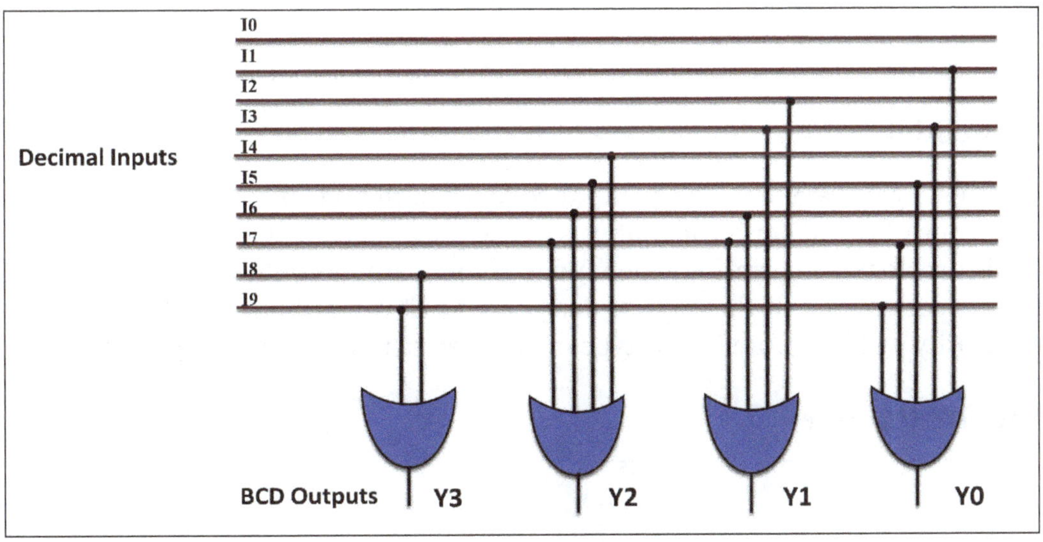

This decimal to BCD encoder design also provides few drawbacks as mentioned in Octal to Binary encoder. The solution to this problem is to use priority encoder and use of additional valid output in the encoder design.

Hexadecimal to Binary Encoder

It accepts sixteen inputs (0, 1, 2...9, A, B, C, D, E, F) and produces 4-bit binary code corresponding to the activated Hexadecimal input. The logic diagram for Hexadecimal to Binary encoder can be developed on the similar line of Decimal to BCD encoder. The truth table for the same encoder can be prepared by simply writing four bit binary code for each activated hexadecimal input as shown in Table.

Table: Truth table for Hexadecimal to Binary encoder.

Hexadecimal Inputs	Binary Outputs			
	Y3	Y2	Y1	Y0
0	0	0	0	0
1	0	0	0	1
2	0	0	1	0
3	0	0	1	1
4	0	1	0	0
5	0	1	0	1
6	0	1	1	0
7	0	1	1	1
8	0	0	0	0
9	1	0	0	1
A	1	0	1	0
B	1	0	1	1
C	1	1	0	0
D	1	1	0	1
E	1	1	1	0
F	1	1	1	1

Priority Encoder

Standard encoders generate the wrong output code when there is more than one input present at logic 1. This is one of the major disadvantages of the linear encoders. One of the simplest ways to overcome this problem is to prioritize the inputs. If there are more than one input at logic '1' then the actual code correspond to the highest designated priority are produced and all other inputs with lower priority will be ignored. This type of encoder is known as priority encoder.

4 to 2 Priority Encoder

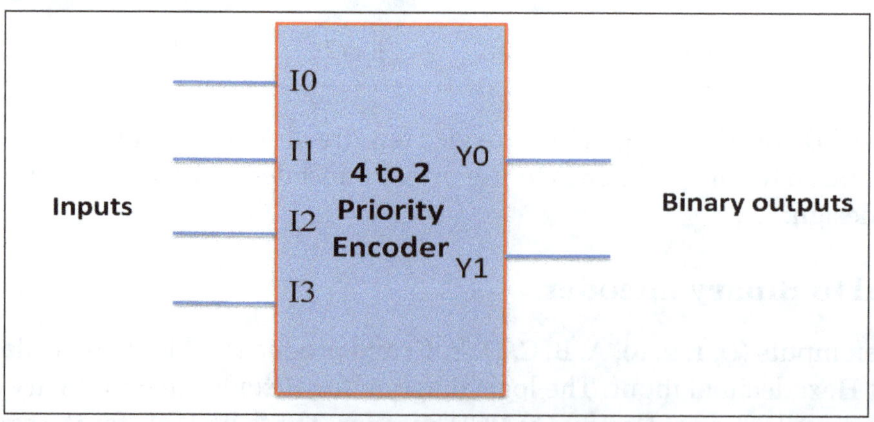

Figure shows the logic symbol of 4 to 2 priority encoder. Priorities are given to the input lines. In this case input I3 is having highest priority than other inputs, I2 has second highest priority as compare to I1 and I0 and so on. If two or more inputs are at logic '1' at the same time, the input line with highest priority will be considered for generating binary output.

Table: Truth table of 4 to 2 priority encoder.

Inputs				Output	
I3	I2	I1	I0	Y1	Y0
0	0	0	0	X	X
0	0	0	1	0	0
0	0	1	X	0	1
0	1	X	X	1	0
1	X	X	X	1	1

There are four inputs I0 through I3 and two outputs Y0 and Y1. With these four inputs, the actual truth table would have been with 16 possible input conditions. But for simplicity a truth table is prepared wisely and contain only few conditions with the help of X – do not care condition. As shown in table. Out of the four inputs I3 has highest priority and I0 has lowest priority. That means when I3=1 then Y1 Y0 = 11 irrespective of the other inputs.

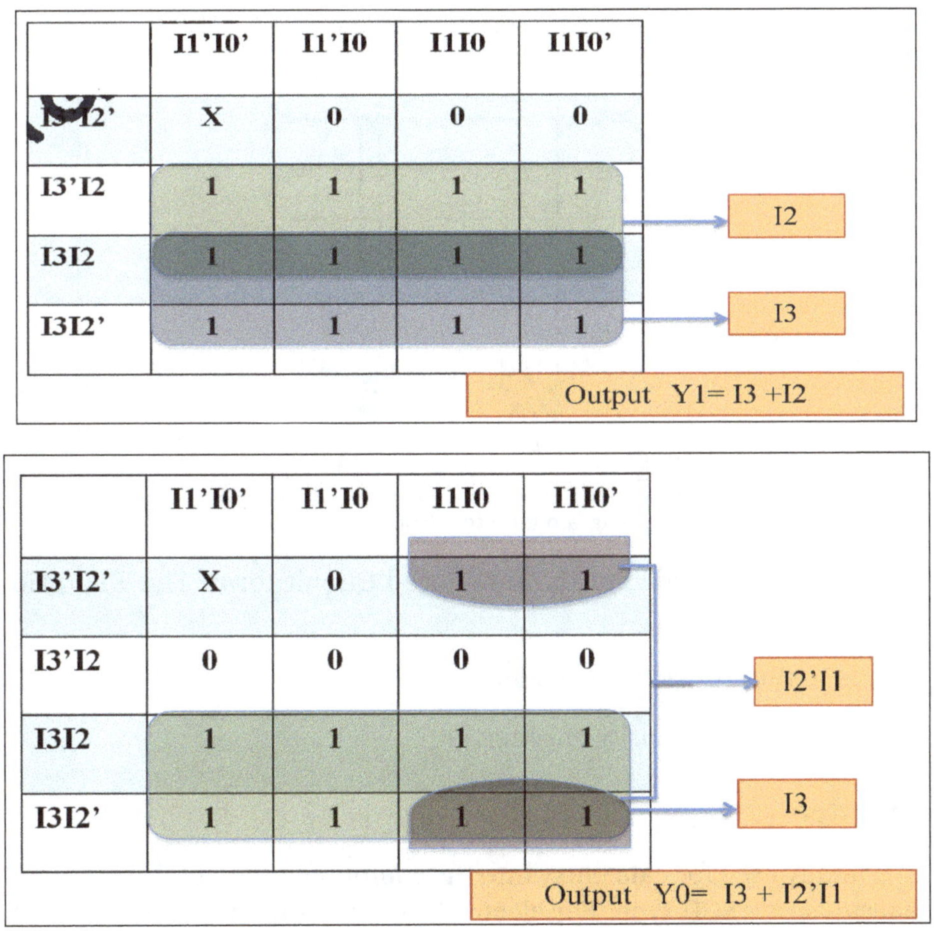

K-maps for the outputs of 4 to 2 line encoder.

The k-maps for two outputs Y1 and Y0 are shown in figure. The logic '1' is mapped for each output for all input conditions. For output Y1, it is possible to group 8 neighboring cells to form the octet. It is possible to form two octets. From these octets, we can eliminate three variables each. The remaining unchanged variables are written in the simplified for. Thus,

$$Output\ Y1 = I3 + I2$$

Similarly we can obtain Boolean expression for output Y0 using another K-map. After mapping the '1's in the map, it is possible to group neighboring cells to form Octet and Quad. Thus, we get,

$$Output\ Y0 = I3 + I2'I1$$

Using these Boolean expressions, it is possible to obtain the logic diagram for 4 to 2 priority encoder. The two binary outputs are obtained with the help of two OR gates, one AND gate and a NOT gate. The logic diagram for 4 to 2 priority encoder is indicated in figure.

If more than one input is active, the higher order input has priority over the lower order input. The higher value is encoded at the output. Further, one can introduce one more output for the priority encoder to indicate whether the output is valid or not. This output is also called as validity indicator. It is generated by simply ORing all possible inputs. If all the inputs are '0', the output is invalid. The output is valid (logic-1) only when at least one input is active.

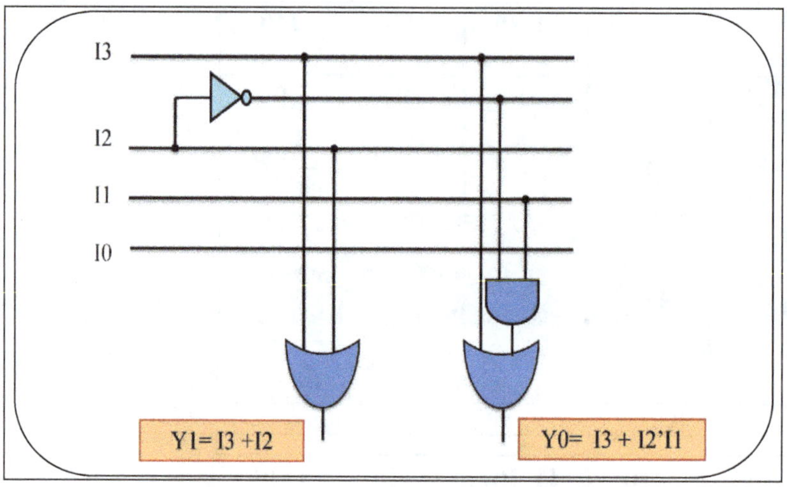

Logic diagram for 4 to 2 Priority Indicator.

Priority Encoders are also available in the Integrated Circuit form. The TTL priority encoders are:

IC 74148 – Octal to Binary priority encoder.

IC 74147 – Decimal to BCD priority encoder.

Keyboard Encoder

Keyboards are generally used for data entry into the computer and digital systems. The Keyboards are either hexadecimal or ASCII type. For electronic calculators there are 10 switches (Keys) for decimal number 0 to 9 and some function keys. When a digit key is pressed the internal circuit understands which key is pressed and Keyboard encode will give a code corresponding to key pressed. A typical Keyboard encoder using IC 74147 a priority Encoder is shown in below figure.

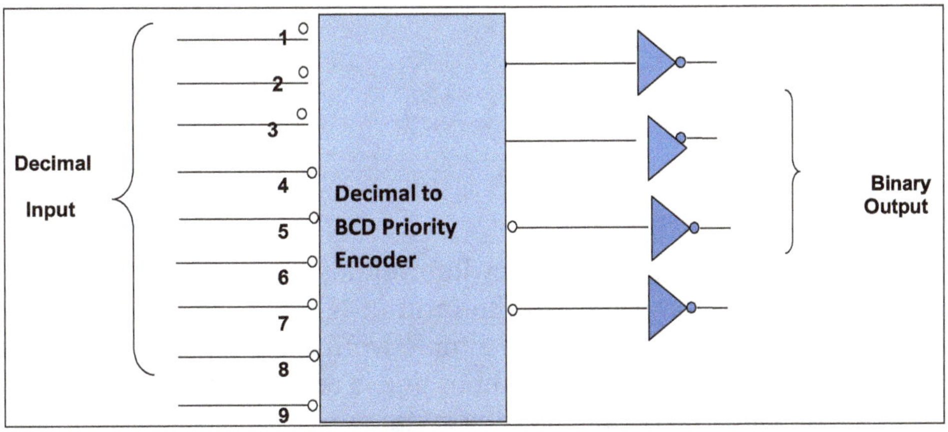

Logic Symbol of Decimal to BCD Priority encoder IC 74147.

There are 10 Push button switches with pull-up resistors connected to the positive Supply. The pulls up resistor make the input line high when the switch is not closed. When a switch is closed the corresponding line will be connected to ground. This provides low level to the corresponding input to the encoder. When key corresponding to decimal digit say 3 is pressed, the encoder IC generates a BCD code of 3 in complemented form. The generated binary code can be converted to proper BCD by using inverter at the output of the encoder. In this encoder, there are nine inputs with highest priority to number 9 and least to number 1.

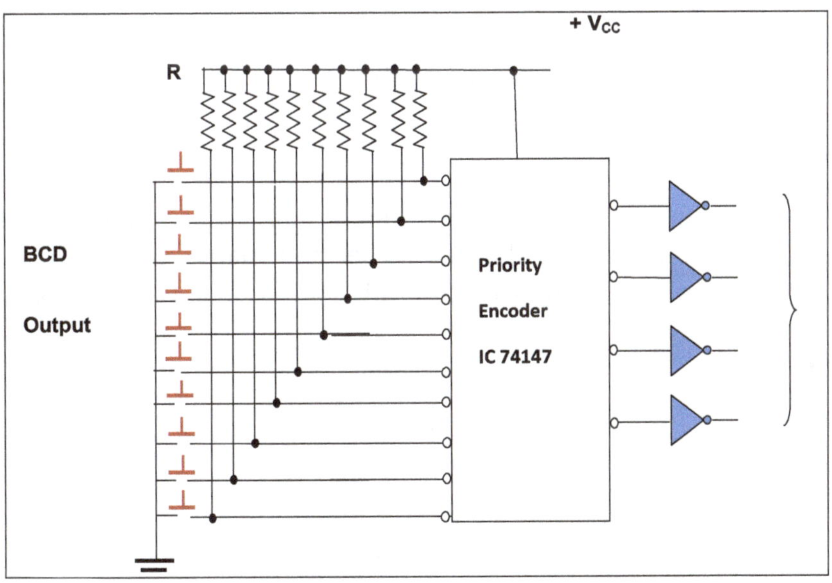

Applications of Encoder

Encoders are normally preferred where there is a need to convert information of code from one format to another. In addition to linear and priority encoders, encoders have many applications as listed below:

- Keyboard encoder for computers,

- Optical encoders –linear or rotary,
- Interfacing peripherals to microprocessors,
- Audio/video coding and transmission.

Decoder

The combinational circuit that change the binary information into 2^N output lines is known as Decoders. The binary information is passed in the form of N input lines. The output lines define the 2^N-bit code for the binary information. In simple words, the Decoder performs the reverse operation of the Encoder. At a time, only one input line is activated for simplicity. The produced 2^N-bit output code is equivalent to the binary information.

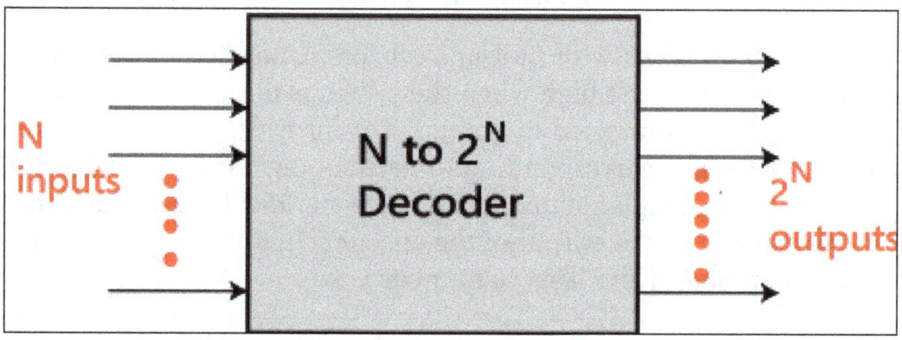

There are various types of decoders which are as follows:

2 to 4 Line Decoder

In the 2 to 4 line decoder, there is a total of three inputs, i.e., A_0, and A_1 and E and four outputs, i.e., Y_0, Y_1, Y_2, and Y_3. For each combination of inputs, when the enable 'E' is set to 1, one of these four outputs will be 1. The block diagram and the truth table of the 2 to 4 line decoder are given below.

Block Diagram

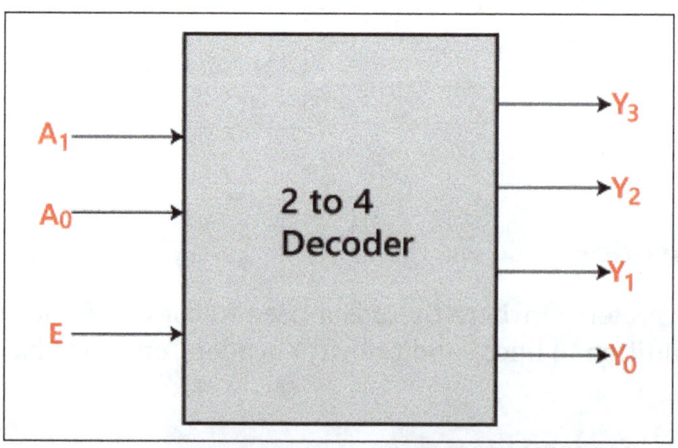

Truth Table

Enable	INPUTS		OUTPUTS			
E	A_1	A_0	Y_3	Y_2	Y_1	Y_0
0	X	X	0	0	0	0
1	0	0	0	0	0	1
1	0	1	0	0	1	0
1	1	0	0	1	0	0
1	1	1	1	0	0	0

The logical expression of the term Y0, Y0, Y2, and Y3 is as follows:

$Y_3=E.A_1.A_0$

$Y_2=E.A_1.A_0'$

$Y_1=E.A_1'.A_0$

$Y0=E.A_1'.A_0'$

Logical circuit of the above expressions is given below:

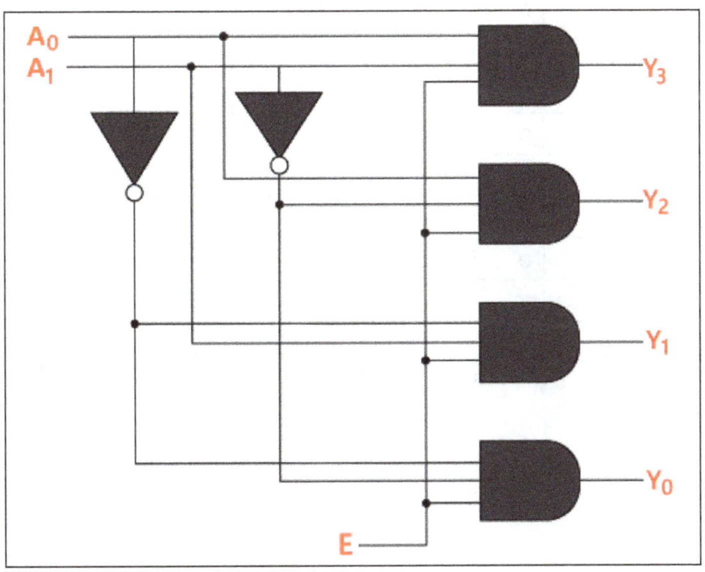

3 to 8 Line Decoder

The 3 to 8 line decoder is also known as binary to Octal Decoder. In a 3 to 8 line decoder, there is a total of eight outputs, i.e., Y_0, Y_1, Y_2, Y_3, Y_4, Y_5, Y_6, and Y_7 and three outputs, i.e., A_0, A1, and A_2. This circuit has an enable input 'E'. Just like 2 to 4 line decoder, when enable 'E' is set to 1, one of these four outputs will be 1. The block diagram and the truth table of the 3 to 8 line encoder are given.

Block Diagram

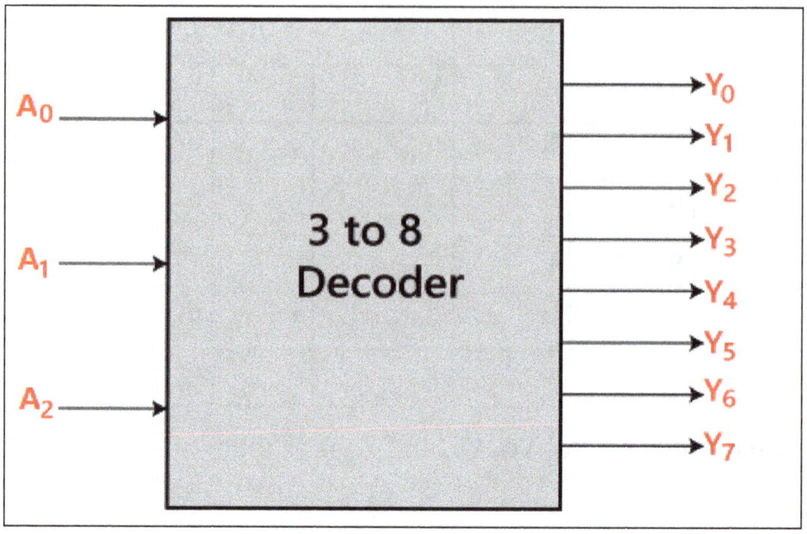

Truth Table

Enable	INPUTS		OUTPUT								
E	A_2	A_1	A_0	Y_7	Y_6	Y_5	Y_4	Y_3	Y_2	Y_1	Y_0
0	x	x	X	0	0	0	0	0	0	0	0
1	0	0	0	0	0	0	0	0	0	0	1
1	0	0	1	0	0	0	0	0	0	1	0
1	0	1	0	0	0	0	0	0	1	0	0
1	0	1	1	0	0	0	0	1	0	0	0
1	1	0	0	0	0	0	1	0	0	0	0
1	1	0	1	0	0	1	0	0	0	0	0
1	1	1	0	0	1	0	0	0	0	0	0
1	1	1	1	1	0	0	0	0	0	0	0

The logical expression of the term Y_0, Y_1, Y_2, Y_3, Y_4, Y_5, Y_6, and Y_7 is as follows:

$Y_0=A_0'.A_1'.A_2'$

$Y_1=A_0.A_1'.A_2'$

$Y_2=A_0'.A_1.A_2'$

$Y_3=A_0.A_1.A_2'$

$Y_4=A_0'.A_1'.A_2$

$Y_5=A_0.A_1'.A_2$

$Y_6=A_0'.A_1.A_2$

$Y_7=A_0.A_1.A_2$

Logical circuit of the above expressions is given below:

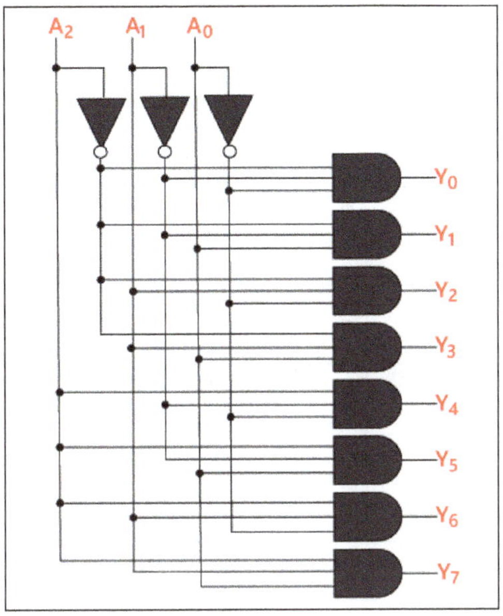

4 to 16 line Decoder

In the 4 to 16 line decoder, there is a total of 16 outputs, i.e., Y_0, Y_1, Y_2,......, Y_{16} and four inputs, i.e., A_0, A1, A_2, and A_3. The 3 to 16 line decoder can be constructed using either 2 to 4 decoder or 3 to 8 decoder. There is the following formula used to find the required number of lower-order decoders.

Required number of lower order decoders=m_2/m_1.

$m_1 = 8$

$m_2 = 16$

Required number of 3 to 8 decoders=$\frac{16}{8}$=2.

Block Diagram

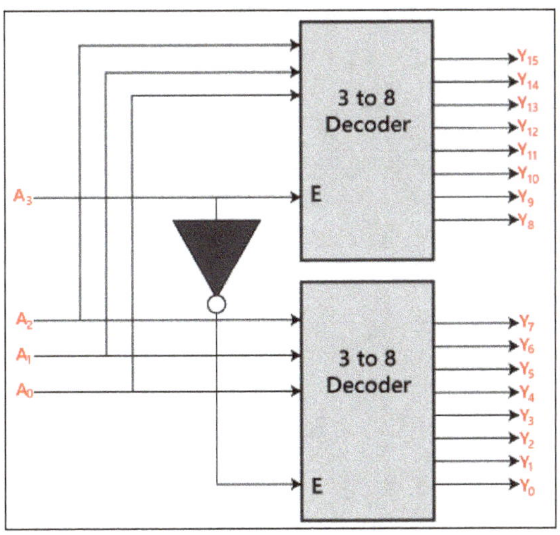

Truth Table

INPUTS				OUTPUTS															
A_3	A_2	A_1	A_0	Y_{15}	Y_{14}	Y_{13}	Y_{12}	Y_{11}	Y_{10}	Y_9	Y_8	Y_7	Y_6	Y_5	Y_4	Y_3	Y_2	Y_1	Y_0
0	0	0	0	0	0	0	0	0	0	0	0	0	0	0	0	0	0	0	1
0	0	0	1	0	0	0	0	0	0	0	0	0	0	0	0	0	0	1	0
0	0	1	0	0	0	0	0	0	0	0	0	0	0	0	0	0	1	0	0
0	0	1	1	0	0	0	0	0	0	0	0	0	0	0	0	1	0	0	0
0	1	0	0	0	0	0	0	0	0	0	0	0	0	0	1	0	0	0	0
0	1	0	1	0	0	0	0	0	0	0	0	0	0	1	0	0	0	0	0
0	1	1	0	0	0	0	0	0	0	0	0	0	1	0	0	0	0	0	0
0	1	1	1	0	0	0	0	0	0	0	0	1	0	0	0	0	0	0	0
1	0	0	0	0	0	0	0	0	0	0	1	0	0	0	0	0	0	0	0
1	0	0	1	0	0	0	0	0	0	1	0	0	0	0	0	0	0	0	0
1	0	1	0	0	0	0	0	0	1	0	0	0	0	0	0	0	0	0	0
1	0	1	1	0	0	0	0	1	0	0	0	0	0	0	0	0	0	0	0
1	1	0	0	0	0	0	1	0	0	0	0	0	0	0	0	0	0	0	0
1	1	0	1	0	0	1	0	0	0	0	0	0	0	0	0	0	0	0	0
1	1	1	0	0	1	0	0	0	0	0	0	0	0	0	0	0	0	0	0
1	1	1	1	1	0	0	0	0	0	0	0	0	0	0	0	0	0	0	0

The logical expression of the term A0, A1, A2,..., A15 are as follows:

$Y_0=A_0'.A_1'.A_2'.A_3'$

$Y_1=A_0'.A_1'.A_2'.A_3$

$Y_2=A_0'.A_1'.A_2.A_3'$

$Y_3=A_0'.A_1'.A_2.A_3$

$Y_4=A_0'.A_1.A_2'.A_3'$

$Y_5=A_0'.A_1.A_2'.A_3$

$Y_6=A_0'.A_1.A_2.A_3'$

$Y_7=A_0'.A_1.A_2.A_3$

$Y_8=A_0.A_1'.A_2'.A_3'$

$Y_9=A_0.A_1'.A_2'.A_3$

$Y_{10}=A_0.A_1'.A_2.A_3'$

$Y_{11}=A_0.A_1'.A_2.A_3$

$Y_{12}=A_0.A_1.A_2'.A_3'$

$Y_{13}=A_0.A_1.A_2'.A_3$

$Y_{14}=A_0.A_1.A_2.A_3'$

$Y_{15}=A_0.A_1.A_2'.A_3$

Logical circuit of the above expressions is given below:

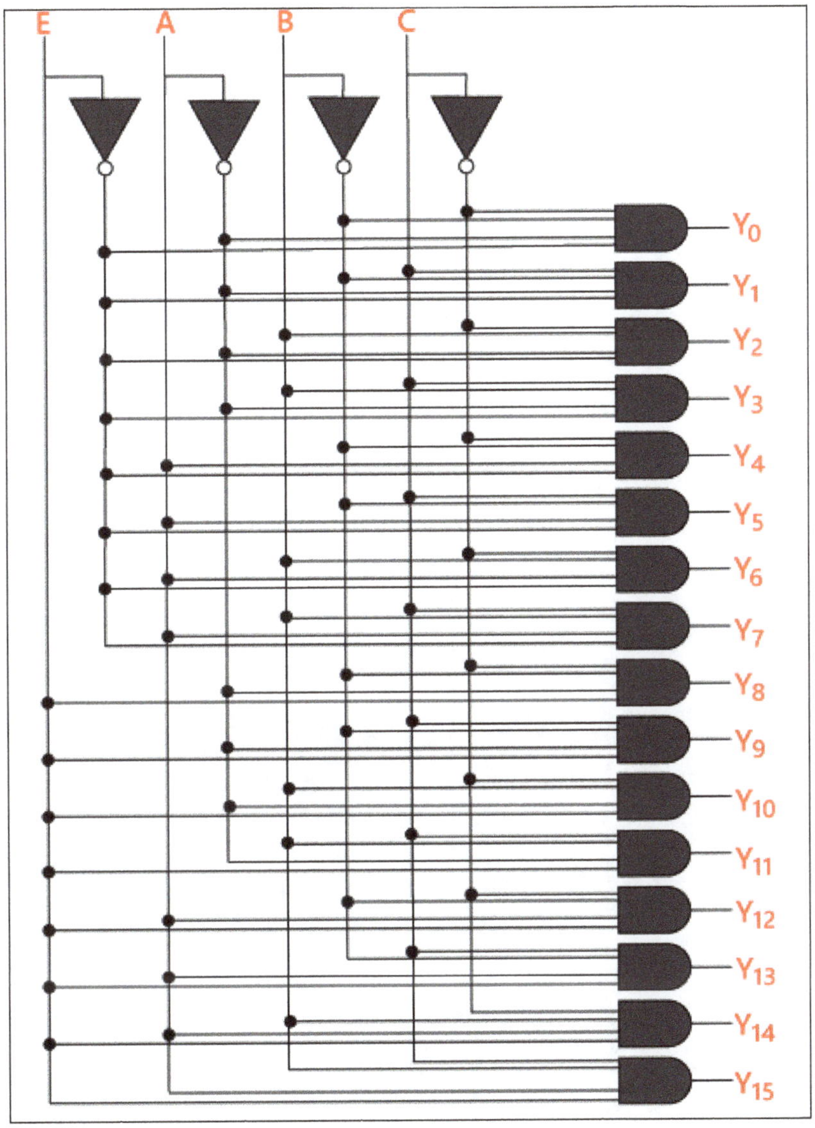

References

- Digital-electronics: electrical4u.com, Retrieved 02, February 2020
- Digital-electronics: electronics.fandom.com, Retrieved 14, August 2020
- Digital-number-system, computer-logical-organization: tutorialspoint.com, Retrieved 09, March 2020
- Boolean-algebra, computer-logical-organization: tutorialspoint.com, Retrieved 12, June 2020
- Decoder-digital-electronics: electronics-tutorials.ws, Retrieved 04, January 2020

3

Power Electronics

Power electronics deals with electronic devices and associated components for the conversion and control of electric power. It makes the use of semiconductor devices like diodes, power transistors and thyristors for conversion. This chapter closely examines the key concepts of power electronics to provide an extensive understanding of the subject.

Power electronics is one of the contemporary subjects of which has seen a lot of advancements in recent times and has impacted human life in almost every sphere. We over selves use so many power electronic applications in our daily life, without even realizing it. Now the question stands as, "what is power electronics?"

We can define power electronics as a subject which is a hybrid of power engineering, analogue electronics, semiconductor devices and control systems. We derive the fundamentals of each subject and apply it in an amalgamated way so as to get a regulated form of electrical energy. Electrical energy in itself is not usable until it is converted into a tangible form of energy such as motion, light, sound, heat etc. In order to regulate these forms of energy, an effective way is to regulate the electrical energy itself and this forms the content of the subject power electronics.

We can trace the overwhelming advancement in the subject back to the development of commercial thyristors or silicon controlled rectifiers (SCR) by General Electric Co. in 1958. Before this the control of electrical energy was mainly done using thyratrons and mercury arc rectifiers which work on the principle of physical phenomena in gases and vapours.

After SCR, a great many power electronic devices have emerged like GTO, IGBT, SIT, MCT, TRIAC, DIAC, IEGT, and IGCT and so on. These devices are rated for several hundreds of volt and ampere unlike the signal level devices which work at few volts and mill amperes. In order to achieve the purpose of power electronics, the devices are made to work as nothing more than a switch. All the power electronic devices act as a switch and have two modes, i.e. ON and OFF.

For example, a BJT (Bipolar Junction Transistor) has three regions of operation in its output characteristics cut-off, active and saturation. In analogue electronics where the BJT is supposed to work as an amplifier, the circuit is so designed to bias it in active region of operation. However in power electronics BJT will work in cutoff region when it is OFF and in saturation region when it is ON.

Now that the devices are required to work as a switch, they must follow the basic characteristic of a switch, i.e. when the switch is ON, it has zero voltage drops across it and carries full current through it, and when it is in OFF condition, it has full voltage drop across it and zero current flowing through it.

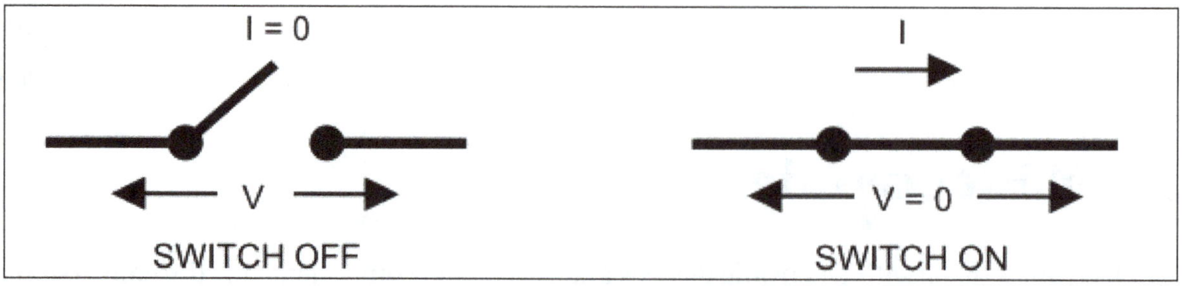

Now since in both the mode either of the quantity V or I is zero, the switch power also turns out to be zero always. This characteristic is easy to visualize in a mechanical switch and the same has to be followed in power electronic switch also.

However practically there always exists a leakage current through the devices when in OFF condition, i.e. $I_{leakage} \neq 0$ and there is always a forward voltage drop in ON condition, i.e. $V_{on} \neq 0$. However the magnitude of V_{on} or $I_{leakage}$ is very less and hence the power across the device is also very less, in order of few mill watts.

This power is dissipated in the device and hence proper heat evacuation from the device is an important aspect. Apart from this ON state and OFF state losses, there are switching losses also in power electronic devices. This is mainly while the switch toggles from one mode to another and V and I across the device changes. In power electronics both the losses are important parameters of any device and essential in determining its voltage and current ratings.

The power electronic devices alone are not that useful in practical applications and hence require to be designed with a circuit along with other supporting components. These supporting components are like the decision making part which controls the power electronic switches in order to achieve the desired output. This includes the firing circuit and the feedback circuit. The block diagram below depicts a simple power electronic system.

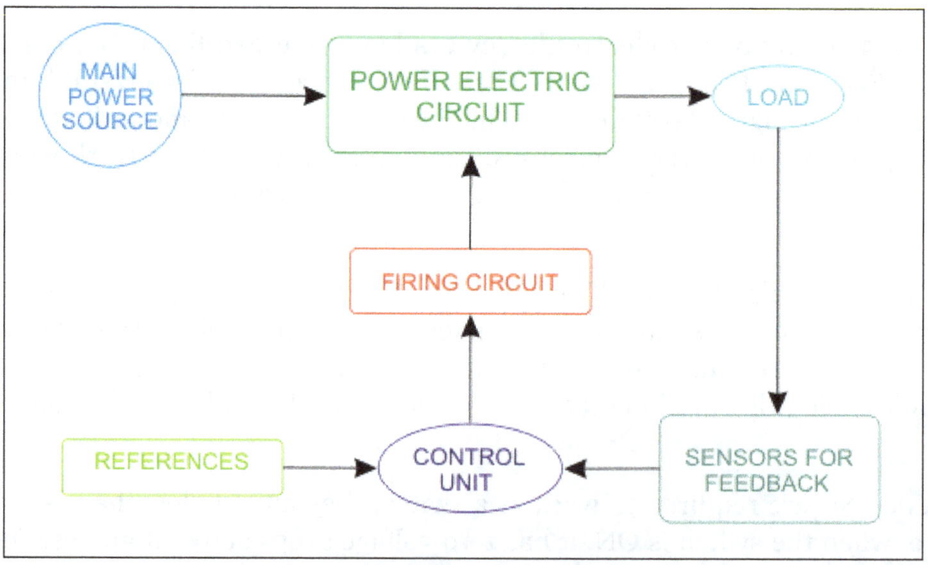

The Control Unit takes the output feedback from sensors and compares it with references and accordingly gives input to the firing circuit. Firing circuit is basically a pulse generating circuit which gives pulse output in a fashion so as to control the power electronic switches in the main circuit block.

The net result is that the load receives the desired electrical power and hence delivers the desired result. A typical example of the above system would be speed control of motors. Majorly there are five types of power electronic circuits, each having different purpose:

- Rectifiers – Converts fixed AC to variable DC (such as half wave rectifiers or full wave rectifiers).
- Choppers – Converts fixed DC to variable DC.
- Inverters – Converts DC to AC having variable amplitude and variable frequency.
- Voltage Regulators – Converts fixed AC to variable AC at same input frequency.
- Cycloconverters – Converts fixed AC to AC with variable frequency.

There is a common misconception about the term converter. Converter is basically any circuit that converts electrical power from one form to another.

Diode and Diode Circuits

A semiconductor diode has two terminals or electrodes that act like an on-off switch. When the diode is "on", it acts as a short circuit and allows current to flow through it. When it is "off", it behaves like an open circuit and passes no current. The two terminals are different and are marked as plus and minus in Figure. If the polarity of the applied voltage matches that of the diode (forward bias), then the diode turns "on".

When the applied voltage polarity is opposite (reverse bias), it turns "off". Of course this is the theoretical behavior of an ideal diode, but it can be seen as a good approximation for a real diode. Diodes are widely used in power electronic circuits for conversion of electric power. The basic application of diode in power electronics is as AC to DC converters commonly known as rectifiers which provide fixed output voltage.

DC and Switching Characteristics of Diode

Diode act as switch to perform different functions such as switches in rectifier, freewheeling diode in switching regulator, charge reversal of capacitor, and energy transfer between components, voltage isolation, energy feedback from the load to the power source, and trapped energy recovery.

Practical diode differs from ideal diode. They have certain limitations. Power diodes are similar to pn-junction signal diodes. However power diode has larger power-, voltage- and current- handling capacities than that of ordinary signal diodes. The frequency response is low compared to signal diode. Practically, power semiconductor diode has $p^{+} n^{-} n^{+}$ structure.

DC Characteristics of Diode

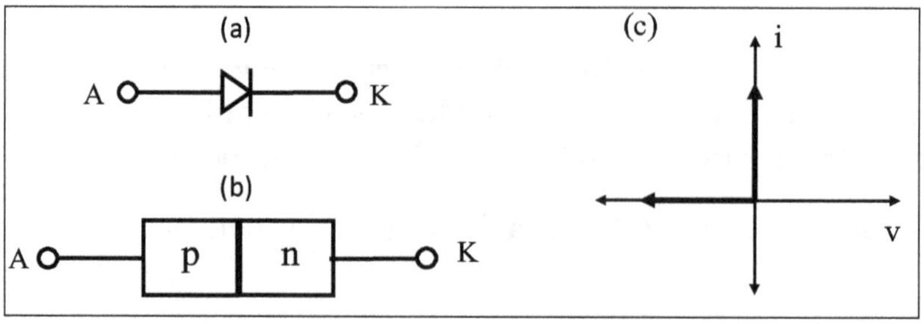

Diode (a) Symbol, (b) structure and (c) ideal current-voltage characteristics.

A power diode is a two terminal pn-junction device. Figure shows the symbol, structure and ideal current voltage characteristics. When anode potential is positive with respect to cathode, the diode is said to be forward biased and the diode conducts. The conduction diode normally has small voltage drop across it. This voltage drop is dependent on the junction temperature and manufacturing process. The diode is said to be in reverse biased when the cathode potential is larger than that of anode. Under reverse biased a small reverse current (called leakage current) flows through it, of the order of few micro or mill-ampere. This reverse current goes on increasing with reverse voltage until avalanche or zener voltage is reached.

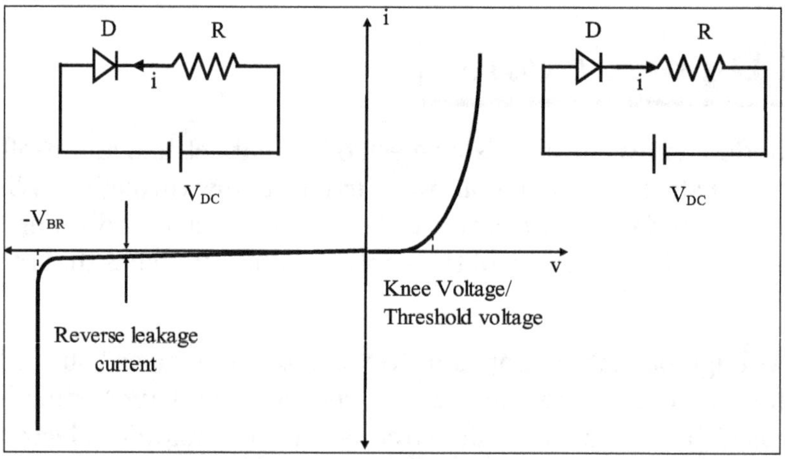

Current-Voltage characteristics of pn-junction diode.

The i-v characteristics of the diode can be expressed by an equation known, as Shockley diode equation, given by: $i = I_s\left(e^{V_D/nV_T} - 1\right)$

Where, i is current through diode,

V_D is Forward biased diode voltage,

I_s is leakage current of the order of 10^{-6} to 10^{-15} A,

V_T is thermal voltage given by $V_T = \dfrac{kT}{q}$.

Where, 'q' is electronic charge (1.6022×10^{-19}C), 'k' is Boltzmann's constant (1.3806×10^{-23}J/K), 'T' is absolute temperature, in Kelvin and 'n' is empirical constant known as emission coefficient or ideality factor, whose value varies between 1 and 2.

Value of 'n' depends on the material and the physical construction of the diode. For Ge diode, 'n' is considered as 1 and Si diode 'n' is 2.

Forward Biased Region: In the forward biased region= $V_D>0$. The diode current ID is very small if the diode voltage is less than the specific value V_{TD} (typically 0.7V for Si diode). The diode conducts fully if $V_D > V_{TD}$, which is referred to as threshold voltage. For small diode voltage V_D=0.1V, $I_D=I_S\, e^{(VD/nV_T)}$.

Reverse Biased Region: In the reverse biased region the VD<0, $I_D \approx -I_S$.

Breakdown region: In the breakdown region the reverse voltage is high (about >1000V). After specific reverse voltage known as V_{BR}, the reverse current increases rapidly with small change in V_{BR}. The operation in the breakdown will not be destructive provided that the power dissipation is within the safe level.

Typical DC Parameters of Diode

- Forward voltage, V_F: It is the voltage drop of a diode across anode and cathode at a defined current level when it is forward biased.
- Breakdown voltage, V_{BR}: It is the voltage drop across the diode at a defined current level when it is beyond reverse biased level. This is popularly known as avalanche breakdown. It is also called as the peak inverse voltage (PIV). The PIV of a diode is the maximum reverse voltage that can be connected across a diode without breakdown. The PIV ratings of power diodes extend from a few volts to several thousand volts.
- Reverse current I_R: I_T is the current at a particular voltage, which is below the breakdown voltage. It is almost constant. It has temperature dependence.

Switching Characteristics of Diode (Reverse Recovery Characteristics)

The current in the forward-biased junction diode is due to net effect of majority and minority charge carriers. Once the diode is in a forward conduction mode and then its forward current is reduced to zero (due to natural behavior of diode circuit or application of reverse voltage), the diode continues to conduct due to minority carriers which remain stored in the pn-junction and the bulk semiconductor material. The minority carriers require a certain time to recombine with opposite charges and to be neutralized. This time is referred to as reverse recovery time (t_{rr}) of the diode. This is as shown in the figure.

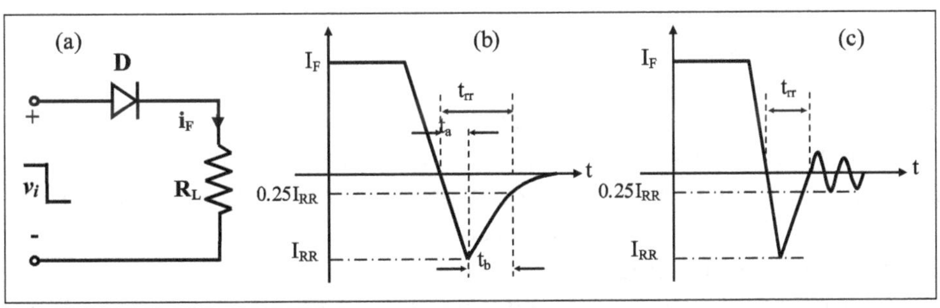

Reverse recovery characteristics showing turn off of a diode (a) circuit arrangement, (b) soft recovery and (c) abrupt recovery.

t_{rr} is measured from the initial zero crossing of the diode current to 25% of the maximum peak current I_{RR}. t_{rr} consists of two components, t_a and t_b as shown in the figure. $t_{rr} = t_a + t_b$, and $I_{RR} = t_a(di/dt)$.

Diodes with soft recovery characteristics are used low frequency applications such as rectifiers and line commutated converters. Diodes with abrupt recovery characteristics are used for high frequency switching applications such as pulse width modulated (pwm) dc-dc converters and inverters.

Typical Switching Parameters of Diode

- Reverse recovery current I_{RR}: It is the peak reverse current, when diode is switched off.
- Forward recovery time, t_{FR}: It is the time required for the diode voltage to drop to a particular value after the forward current starts to flow.
- Reverse recovery time t_{rr}: It is the time interval between the application of reverse voltage and the reverse current dropped to a particular value as shown in Figure.

Parameter t_a is the interval between the zero crossing of the diode current and when it reaches to I_{RR}. On the other hand, t_b is the time interval from the maximum reverse recovery current to approximately $0.25I_{RR}$.

- Softness factor (SF): It is the ratio of t_a and t_b. $SF = t_b / t_a$.

Types of Power Diodes

Ideal diode should have no reverse recovery time. But the manufacturing cost of such diode will increase. So in applications where the effect of reverse recovery time is not significant, then inexpensive diodes can be used. Power diodes can be classified depending on the manufacturing techniques, forward voltage drops and the reverse recovery characteristics of the diode. Following are the categories of power diode:

- Standard/ general purpose diodes,
- Fast recovery diodes,
- Schottky diodes,
- Silicon Carbide diodes.

General Purpose Diodes

These types of diodes have relatively high reverse recovery time, of the order of 25μs. these diodes are used in low speed applications, since in low speed applications, the reverse recovery time is not critical. As an example in diode rectifiers and converters for low frequency up to 1 kHz applications and line commutated applications these diodes can be used. These diodes have current rating less than 1A to a few kA ranges with voltage rating of 50V to about 5kV. The manufacturing process of these diodes is by diffusion technique. Alloyed types of rectifiers which are used in welding applications have voltage/ current rating up to 1.5kV/ 400A.

Fast-recovery Diodes

These types of diodes have low reverse recovery time, less than 5μs. they have applications in dc-ac converter circuits, dc-dc converter circuits. In these applications reverse recovery time is of critical importance. These diodes are made by using diffusion technique for voltage rating above 400V. The reverse recovery time is controlled by diffusion of gold or platinum.

The epitaxial growth technique is used to fabricate the diode which has voltage rating less than 400V. These epitaxial diodes have faster switching speed than those of diffusion diodes. The epitaxial diodes have a narrow base width, which provides faster reverse recovery time as low as 50ns. These fast recovery diodes are available in various sizes.

Schottky Diodes

The pn junction diode has charge storage problem. To minimize/ eliminate this problem Schottky diodes are produced. It is accomplished by setting up a barrier potential with a contact between a metal and a semiconductor. A layer of metal is deposited on a thin epitaxial layer of n-type silicon semiconductor. The potential barrier simulates the behavior of pn junction. The rectifying action depends on the majority charge carriers only. So there are no minority charge carriers to recombine. The recovery effect is due to only the self-capacitance of the semiconductor junction. The recovered charge of the Schottky diode is much less compared to pn junction diode. Since the recovered charge is only due to junction capacitance, it is largely independent of reverse rate of change of current di/dt. This diode has relatively small forward voltage drop. The leakage current of these diodes is more than that of pn junction diode. In Schottky diodes smaller the conduction voltage more is the leakage current. Thus, maximum allowable voltage is limited to 100V. The current rating varies from 1A to 100A.

Silicon Carbide Diodes (SiC Diodes)

SiC (Silicon Carbide) is a compound semiconductor comprised of silicon (Si) and carbon (C). Compared to Si, SiC has ten times the dielectric breakdown field strength, three times the band gap, and three times the thermal conductivity. Both p-type and n-type regions can be formed in SiC. These properties make SiC an attractive material from which to manufacture power devices that can far exceed the performance of their Si counterparts. SiC devices can withstand higher breakdown voltage, have lower resistivity, and can be operated at higher temperature. The larger band gap also means SiC devices can operate at higher temperatures. The guaranteed operating temperature of current SiC devices is from 150°C – 175°C. This is due mainly to thermal reliability of packages. When properly packaged, they can be operated at 200°C and higher.

Device Structure and Characteristics

SiC SBDs (Schottky Barrier Diodes) with breakdown voltage from 600V (which far exceeds the upper limit for silicon SBDs) and up are readily available. Compared to silicon FRDs (fast recovery diodes), SiC SBDs have much lower reverse recovery current and recovery time, hence dramatically lower recovery loss and noise emission. Furthermore, unlike silicon FRDs, these characteristics do not change significantly over current and operating temperature ranges. SiC SBDs allow system designers to improve efficiency, lower cost and size of heat sink, increase switching frequency to reduce size of magnetics and its cost, etc. SiC-SBDs are increasingly applied to circuits such as power factor correctors (PFC) and secondary side bridge rectifier in switching mode power supplies. Today's applications are air conditioners, solar power conditioners, EV chargers, industrial equipment and so on.

Forward Characteristics of SiC-SBD

SiC-SBDs have similar threshold voltage as Si-FRDs, i.e., a little less than 1V. Threshold voltage is determined by Schottky barrier height. Normally, a low barrier height corresponds with low threshold voltage and high reverse leakage current.

Reverse Recovery Characteristics of SiC-SBD

Si fast P-N junction diodes (e.g. FRDs: fast recovery diodes) have high transient current at the moment the junction voltage switches from the forward to the reverse direction, resulting in significant switching loss. This is due to minority carriers stored in the drift layer during conduction phase when forward voltage is applied. The higher the value of forward current (or temperature), the longer the recovery time and the larger is the recovery current. In contrast, since SiC-SBDs are majority carrier (unipolar) devices that use no minority carriers for electrical conduction, they do not store minority carriers. The reverse recovery current in SiC SBDs is only to discharge junction capacitance. Thus the switching loss is substantially lower compared to that in Si-FRDs. The transient current is nearly independent of temperatures and forward currents, and thereby achieves stable fast recovery in any environment.

Type	V_{BR}	V_F	t_{rr}
Si Schottky Barrier Diode	15 V - 200 V	0.3V-0.8 V	< 10 ns
Si Rectifier Diode (General purpose)	50 V – 5kV	0.7V-1.5 V	> 25 μs
Si Standard Recovery Diode	50 V- 1kV	1.0 V	1 μs-5 μs
Silicon Carbide Schottky Barrier Diode	600 V	1.5 V	< 15 ns

Snubber Circuits for Diode

Snubber circuits are essential for diodes used in switching circuits. It can save a diode from over-voltage spikes, which may arise during the reverse recovery process. A very common snubber circuit for a power diode consists of a capacitor in series with a resistor connected in parallel with the diode as shown in Figure.

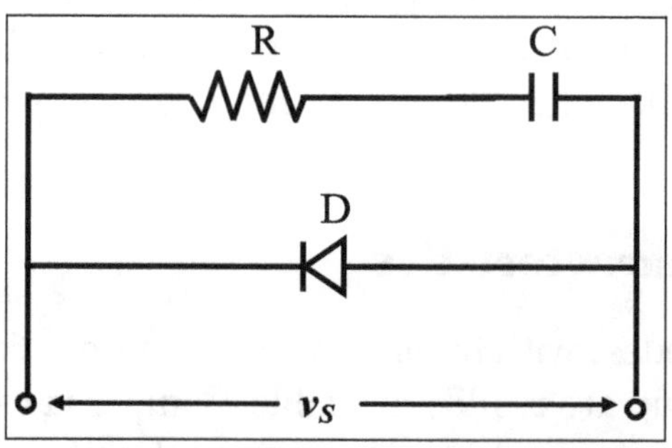

When the reverse recovery current decreases, the capacitor by virtue of its property will try to hold the voltage across it, which, approximately, is the voltage across the diode. The resistor on the other hand will help to dissipate some of the energy stored in the inductor, which forms the I_{RR} loop. The dv/dt across a diode can be calculated as,

$$\frac{dv}{dt} = \frac{0.632V_S}{\tau} = \frac{0.632V_S}{R_S C_S}$$

where V_S is the voltage applied across the diode.

Usually the dv/dt rating of a diode is given in the manufacturer's datasheet. Knowing dv/dt and the R_S, one can choose the value of the snubber capacitor $_{CS.}$ The R_S can be calculated from the diode reverse recovery current,

$$R_S = \frac{V_S}{I_{RR}}$$

The designed dv/dt value must always be equal or lower than the dv/dt value found from the datasheet.

Series Connected Diodes

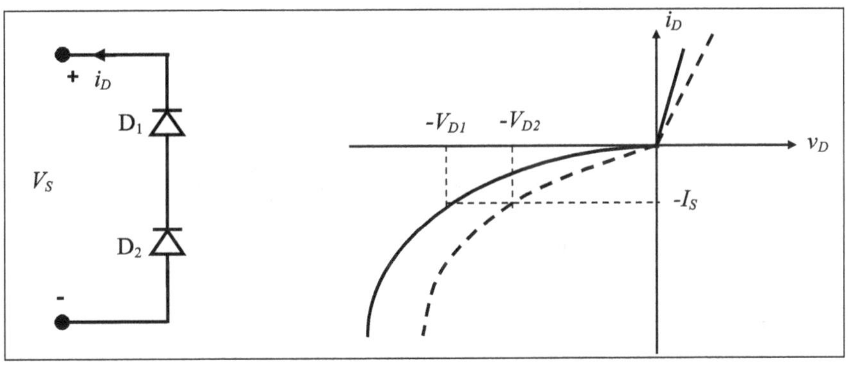

Series connected diode (a) circuit (b) reverse bias characteristics showing unbalanced voltage sharing.

Figure shows series connection of two diodes. If diodes are connected in series as shown there may be unequal voltage sharing during reverse bias. It is due to the dissimilarities between the reverse characteristics of diodes.

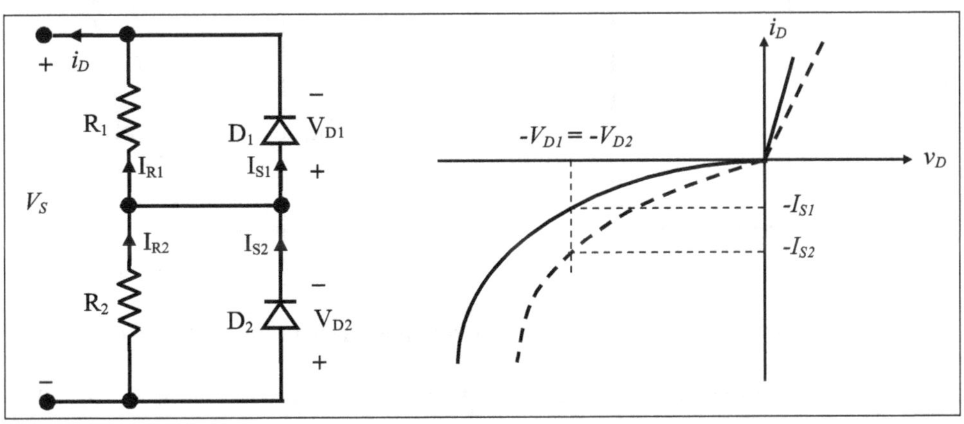

Series connected diode (a) circuit with unbalanced voltage sharing network under steady state condition (b) reverse bias characteristics.

The voltage sharing across diodes can achieved using high value resistors across each diode. Figure shows the unbalanced voltage sharing network under steady state condition with reverse biased characteristics. The current flowing through the resistors should be much greater than the leakage current of diode to ensure the required voltage sharing.

High diode current may flow under voltage transient conditions due to finite reverse recovery time. Snubber circuit is required across each diode to suppress the voltage transients. Figure shows the diode with voltage sharing network under steady state and snubber circuit for transient condition.

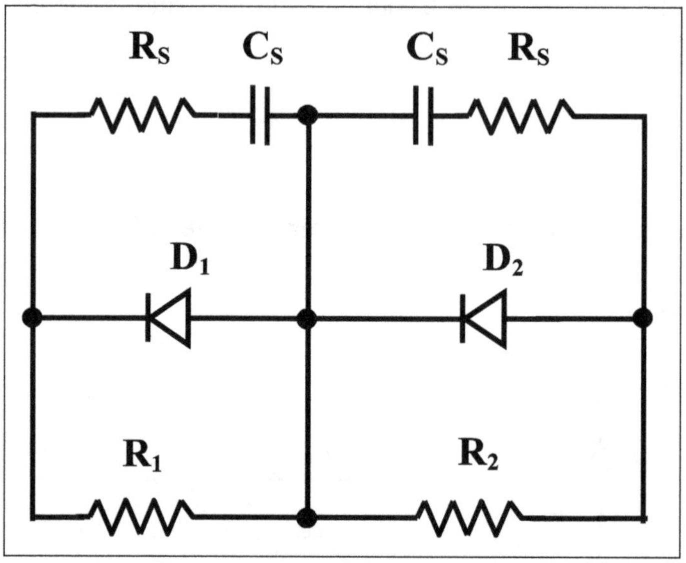

Series connected diode with voltage sharing network under steady state and snubber circuit for transient conditions.

Parallel Connected Diodes

If diodes are connected in parallel, as shown in Figure, current is shared, but current sharing will not be equal because of difference in forward characteristics of diodes.

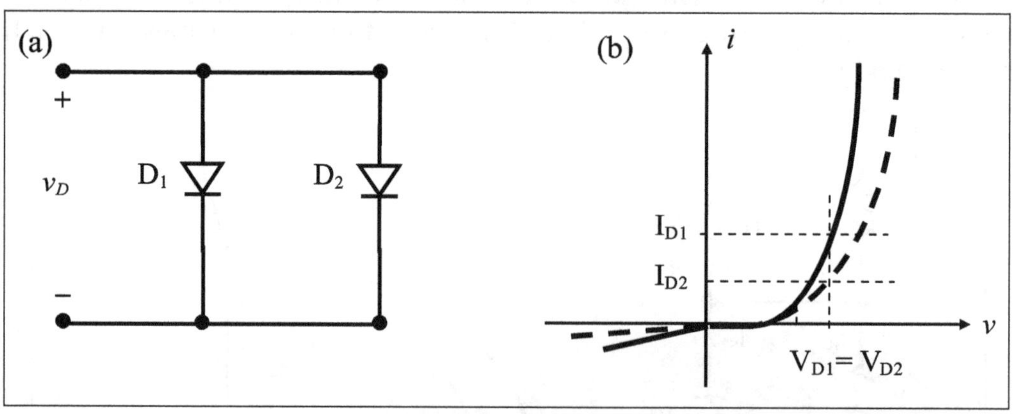

Parallel connected diodes (a) circuit (b) forward biased steady state characteristics.

The series diode with low resistance will improve the current sharing. The circuit arrangement is shown in Figure. If resistor values are high, the sharing will be good, but power dissipation across resistors will be high. It will reduce the efficiency. Thus selection of resistor is difficult.

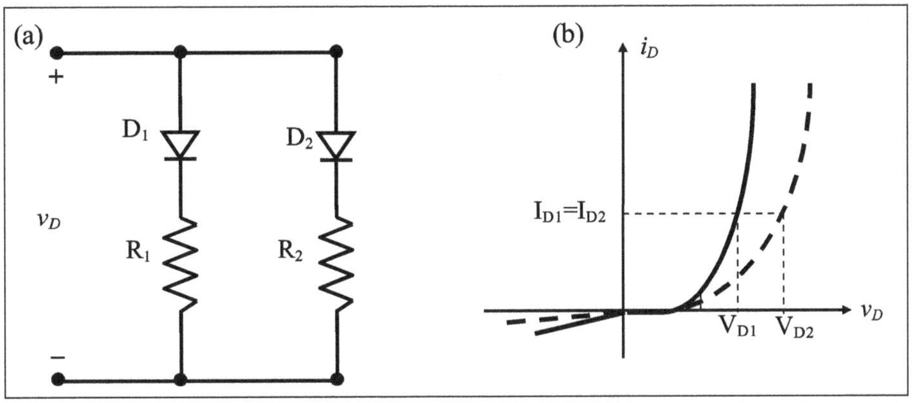

Parallel connected diodes (a) circuit with static current sharing (b) forward biased steady state characteristics.

If low value resistor is selected, dynamic current sharing as shown in Figure can be used. A series coupled reactor can be used. The series inductors, L_1 and L_2 are connected in out of phase. Therefore dynamic current sharing is possible. Time required for current sharing will be dependent on the L/R time constant.

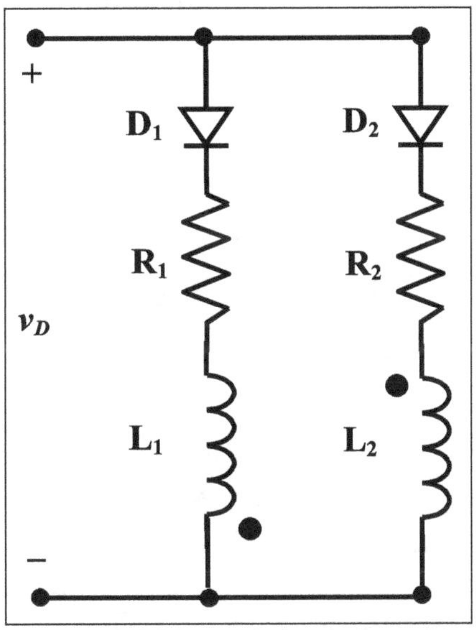

Parallel connected diodes with dynamic current sharing.

Behavior of Diode Circuits

Assumptions

The analysis of diode circuit usually begins with the study of the idealized version of the circuit. Practical deviations from the idealized circuit can then be added whenever needed. The simplifying assumptions involved in the idealized circuits are as follows:

- The voltage drop across switching devices is neglected while they are conducting, and the leakage current is neglected while they are blocking.
- The turn-on and turn-off times of the switching devices are negligible and hence neglected.
- Sources exhibits zero source impedance.

Diode with Series RC Loads

Figure shows the diode circuit for RC load. When switch S1 closed at t=0, the charging current i is found from the following equation:

$$V_S = v_R + v_c = v_R + \frac{1}{C}\int idt + v_c(t=0)$$

$$v_R = R$$

charging current is given by: $i(t) = \frac{V_S}{R} e^{-t/RC}$,

the capacitor voltage is given by, $v_c(t) + \frac{1}{C}\int_0^t idt = V_s\left(1-e^{-t/RC}\right) = V_s\left(1-e^{-t/\tau}\right)$,

where τ=RC is time constant of the RC load. The rate of change of the capacitor voltage is given by:

$\frac{dv_c}{dt} = \frac{V_s}{RC} e^{-t/RC}$. The initial rate of change of capacitor voltage is obtained from above equation as,

$$\left(\frac{dv_c}{dt}\right)_{t=0} = \frac{V_s}{RC}.$$

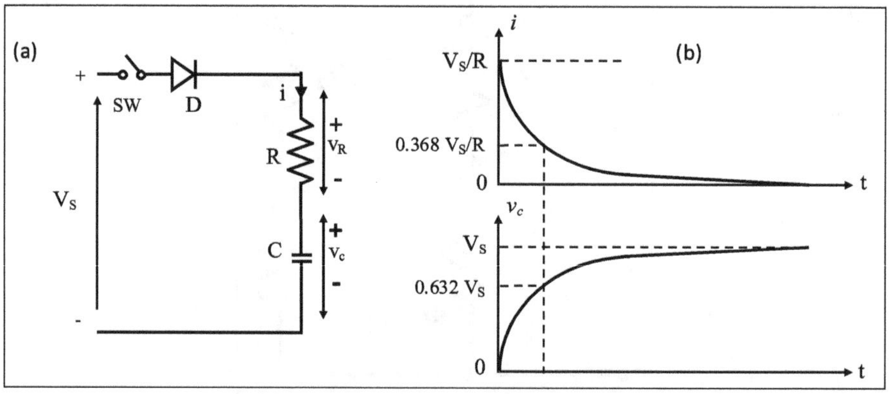

Diode with series RC load (a) circuit, (b) timing waveform.

Diode with Series LR Loads

Figure shows the diode circuit for LR load. When switch is closed at t=0, the current i through the inductor increases and given by the equation,

$$V_s = v_L + v_R = L\frac{di}{dt} + Ri$$

With initial condition i (t=0) =0, the solution of the equation becomes,

$$i(t) = \frac{V_s}{L}\left(1-e^{-tR/L}\right)$$

The rate of change of this current is given by,

$$\frac{di}{dt} = \frac{V_s}{L} e^{-tR/L}$$

And the initial rate of rise of current (at t=0) is obtained as,

$$\left(\frac{di}{dt}\right)_{t=0} = \frac{V_s}{L}$$

The voltage v_L across the inductor is,

$$v_L(t) = L\frac{di}{dt} = v_s e^{-tR/L}$$

Where, $\frac{L}{R} = \tau$ is the time constant of the R-L load.

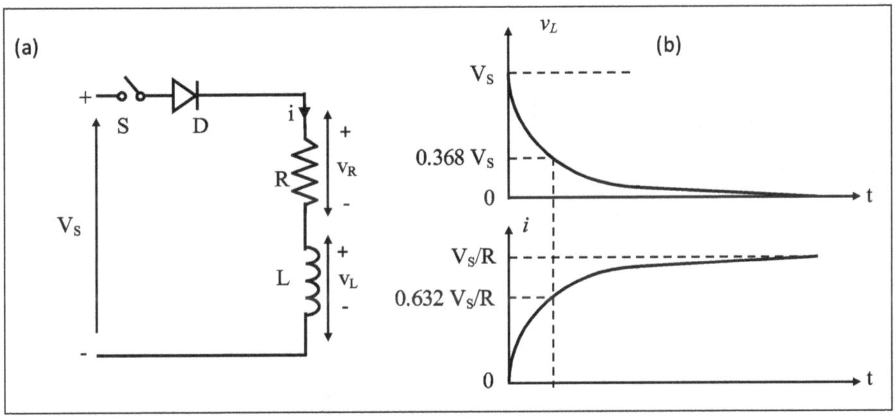

Diode with series R-L load (a) circuit, (b) timing waveform.

Diode with Series LC Load

A diode with LC load is as shown in the Figure.

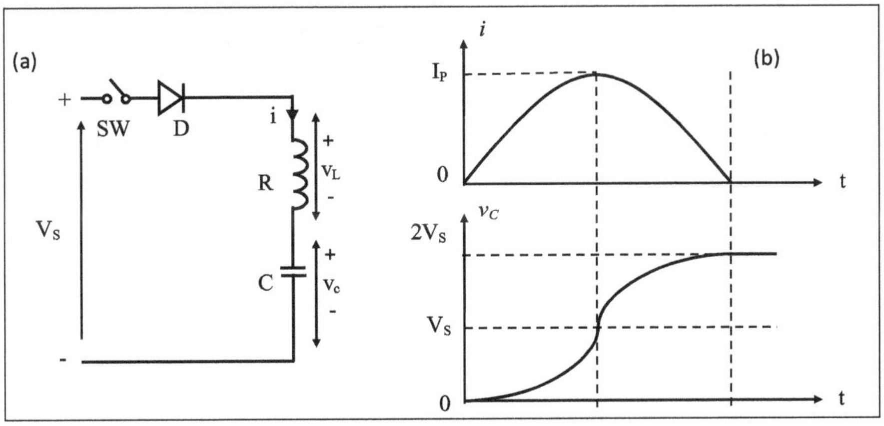

Diode with series LC load (a) circuit, (b) timing waveform.

When the switch S_1 is closed at t=0, the charging current i of the capacitor is expressed as:

$$V_s = L\frac{di}{dt} + \frac{1}{C}\int idt + v_c(t=0)$$

With initial conditions $i(t=0)=0$ *and* $v_c(t=0)=0, i(t) = V_s\sqrt{\frac{C}{L}}\sin\omega t$

$$\therefore i(t) = I_p \sin\omega t$$

Where, $\omega = 1/\sqrt{LC}$ *and*,

The peak current I_p is, $I_p = V_s\sqrt{\frac{C}{L}}$,

The rate of rise of current is obtained from: $\frac{di}{dt} = \frac{V_s}{L}\cos\omega t$,

The voltage across the capacitor can be derived as: $v_c(t) = \frac{1}{C}\int_0^t idt = V_s(1-\cos\omega t)$,

At time $t = t_1 = \pi\sqrt{LC}$ the diode current falls to zero and the capacitor is charged to $2V_s$. Time constant s given by $\tau = \sqrt{LC}$.

The Diode with Series R-L-C Load

A diode with R-L-C load is as shown in the Figure.

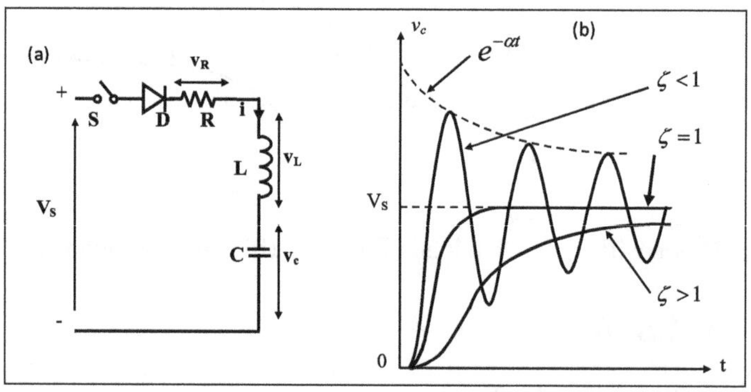

Diode with series R-L-C load (a) circuit, (b) timing waveform.

If switch S_1 is closed at time t=0, the equation using KVL to the circuit can be given by:

$$L\frac{di}{dt} + Ri + \frac{1}{C}\int idt + v_c(t=0) = v_s$$

With initial conditions, i(t=0) = V_0. Differentiating above equation and rearranging the terms we get:

$$\frac{d^2i}{dt} + \frac{R}{L}\frac{di}{dt} + \frac{i}{LC} = 0$$

Under steady state conditions, the capacitor is charged to the source voltage V_s and the steady state current will be zero. The force component of the current will also be zero. The current is due to natural component.

The characteristics equation in Laplace domain can be written as: $s^2 + \frac{R}{L}s + \frac{1}{LC} = 0$,

And the roots of quadratic equation can be written as,

$$s_{1,2} = -\frac{R}{2L} \pm \sqrt{\left(\frac{R}{2L}\right)^2 - \frac{1}{LC}} = -\alpha \pm \sqrt{\alpha^2 - \omega_0^2},\ where\ \alpha = \frac{R}{2L}\ and\ \omega_0 = \frac{1}{\sqrt{LC}}.$$

The term α is a damping attenuation or damping factor and ω_0 is natural or resonating frequency. The ratio of $\zeta = \alpha/\omega_0$ is called as damping ratio. It is evident that the solution for the current depends on damping factor. Hence, there will be three cases for current as follows:

Case 1

$\zeta = 1$ i.e. $\alpha = \omega_0$, the roots are real. Both roots are equal, $s_1 = s_2$. This condition is said to be critically damped circuit. Solution of the equation is,

$$i(t) = (A_1 + A_2 t)e^{s_1 t}$$

Case 2

$\zeta > 1$ i.e. $\alpha > \omega o$, the roots are real. This condition is said to be over damped circuit. Solution of the equation is,

$$i(t) = A_1 e^{s_1 t} + A_2 e^{s_2 t}$$

Case 3

$\zeta < 1$ i.e. $\alpha < \omega_0$, the roots are complex. This condition is said to be under damped circuit. The roots are $s_{1,2} = -\alpha \pm j\omega_r$, where ω_r is called as ringing or damped resonant frequency and is equal to $\omega_r = \sqrt{\omega_0^2 - \alpha^2}$. The solution will be damped or decaying sinusoidal given as follows. Solution of the equation is,

$$i(t) = e^{-at}(A_1 \cos \omega_r t + A_2 \sin \omega_r t)$$

The waveforms of the capacitor voltage are as shown in above Figure.

Freewheeling Diodes

Practically, circuit will produce high voltage transient when circuit is turned off before inductor current reaches to zero. Abrupt change in inductor current while turning off a diode produces the high negative voltage transient. If the transient voltage exceeds breakdown voltage of diode D_1, diode may damage. To avoid such situation additional diode is used as shown in Figure. This diode D_m is called as freewheeling diode. This diode D_m is also described as commutating / flywheel / bypass diode. The main function of this diode is to commutate or transfer load current away whenever load voltage reverses.

Figure shows the diode circuit connected to load with additional freewheeling diode D_m. If the switch S_1 of the diode circuit for R-L load is closed for time t_1, a current is established through the load; and then if the switch is opened, a path must be provided for the current in the inductive load. This is normally done by connecting the diode D_m as shown in the Figure, which is referred to as freewheeling diode. The circuit operation is divided in two modes as described below.

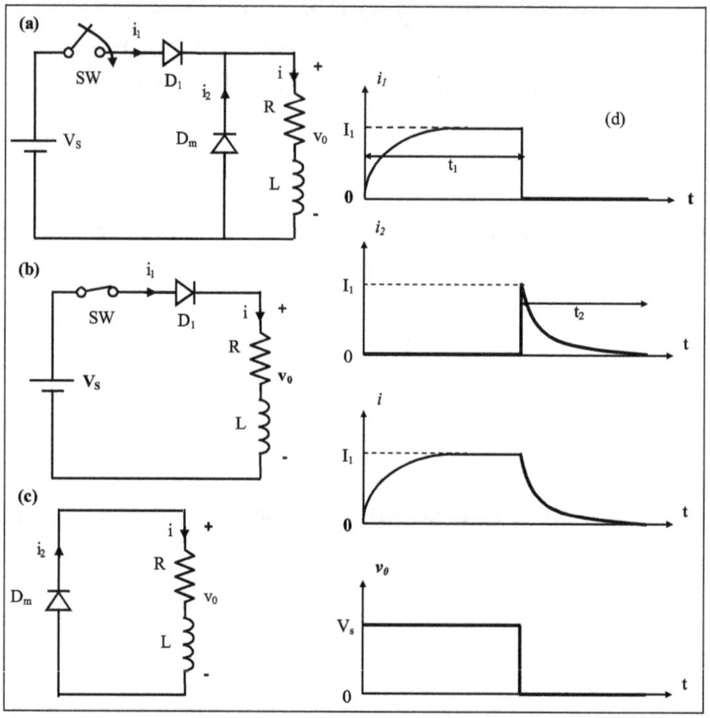

Diode connected to series load with freewheeling diode (a) circuit, (b) equivalent circuit during mode 1, (c) equivalent circuit during mode 2 and (d) timing waveforms.

1.) Mode 1: During this mode, Switch is closed at time t = 0. Assume the initial current at t = 0 is zero. The switch is closed till time $t = t_1$. The equivalent electrical circuit during this mode is shown in Figure and the various currents flowing through D_1, D_m and load and load voltage is shown using timing diagram in Figure.

The current i_1, during this mode can be determined and is given by,

$$i_1(t) = \frac{V_s}{R}\left(1 - e^{-t_1 R/L}\right)$$

When the switch is opened at time $t=t_1$ (at the end of this mode), the current at that time becomes,

$$I_1 = i_1(t = t_1) = \frac{V_s}{R}\left(1 - e^{-t_1 R/L}\right).$$

If the time t_1 is sufficiently long, the current reaches to steady state value and a steady state current of $I_s = V_s / R$ flows through the load.

2.) Mode 2: When switch is opened at time $t = t_1$, the load current starts to flow through the freewheeling diode D_m as shown in the timing waveform. The initial condition of current i_2 is $i_2(t = t_1) = I_1$. Using KVL, we can write,

$$0 = L\frac{di_2}{dt} + Ri_2$$

The solution gives the freewheeling current $i_2(t) = I_1 e^{-t_1 R/L}$. This current decays exponentially to zero at $t = t_2$ provided that $t_2 \gg L/R$.

Characteristics of PN Junction Diode

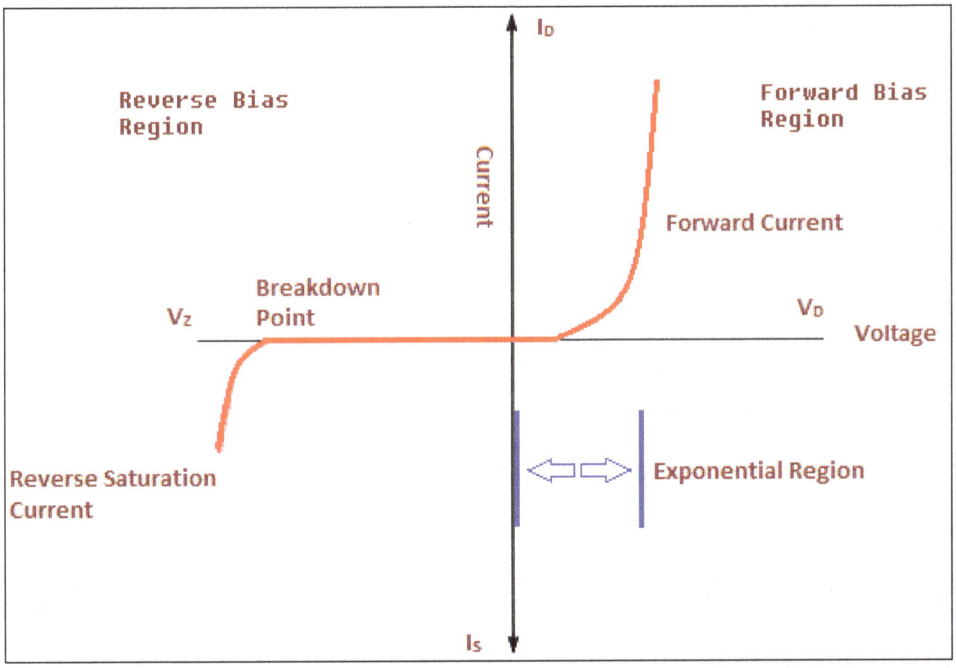

The VI characteristics of a PN Junction Diodes can be categorized into two parts i.e. VI characteristics during Forward Bias and Reverse Bias. The figure shows the Voltage-Current characteristics in diodes.

Characteristics of Diode in Forward Bias

The non–linear curve indicates that when the p-n junction is forward biased, the electrical resistance, impedance is low and conducts a large amount of current known as infinite current. In forward bias region, the VI characteristic is explained with the equation:

$$I_D = I_s\left(e^{\frac{v_D}{nv_T}} - 1\right)$$

Here, $V_T = KT/q$

Where,

- I_D = Forward current,
- I_s = Saturation current,
- V_T = Forward voltage,
- $K = 1.38 \times 10^{-23}$(Boltzmann's constant),
- T = Absolute temperature,
- $q = 1.6 \times 10^{-19}$ (electronic charge),
- n = a constant and has a value of 1 or 2 which depends on the material and physical structure of the diodes:
    - n=1 for Silicon and Germanium,
    - n=2 for Gallium Arsenide.

Characteristics of Diode in Reverse Bias

The current in reverse bias is low till breakdown is reached and hence diode looks like an open circuit. When the input voltage reaches breakdown voltage, reverse current increases enormously. The current value ($I_D = -I_S$) is so small that we can approximate it to zero. Hence we can Say that:

- $I_D \approx 0$ if $-V_Z K < V_D << -n V_T$,
- $\therefore V_D \approx 0.7$ if $I_D > 0$ (Forward bias),
- $I_D \approx 0$ if $-V_Z K < V_D < 0$ (Reverse bias).

Applications of Diode

Diodes are mainly used in rectifiers, clippers, clampers, snubber circuits and voltage regulators.

Use of Diode as Rectifier

Conversion of AC input voltage to the DC output voltage when AC voltage is applied to a diode is called as rectification. Diodes are used individually or connected together to produce various rectifier circuits such as half-wave and full-wave rectifiers for power and signal rectification.

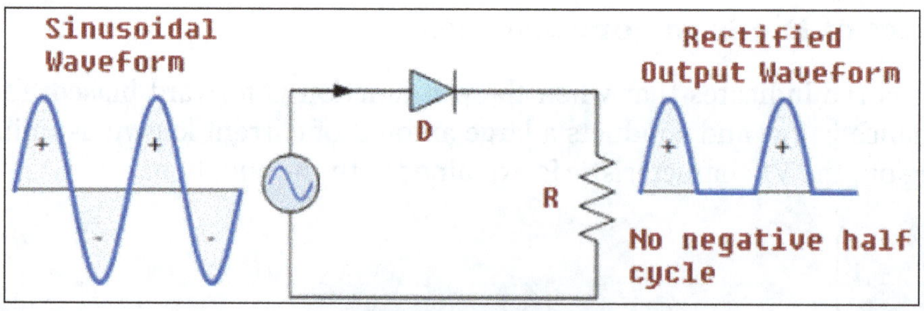

Diode Rectifier Circuit.

Use of Diode as Clipper

Clipping is a waveform shaping where the input signal is clipped or cut to produce an output which is a flattened version of the input signal. Diode-clipping circuits are used in voltage limiting applications as this circuit eliminates voltages below zero.

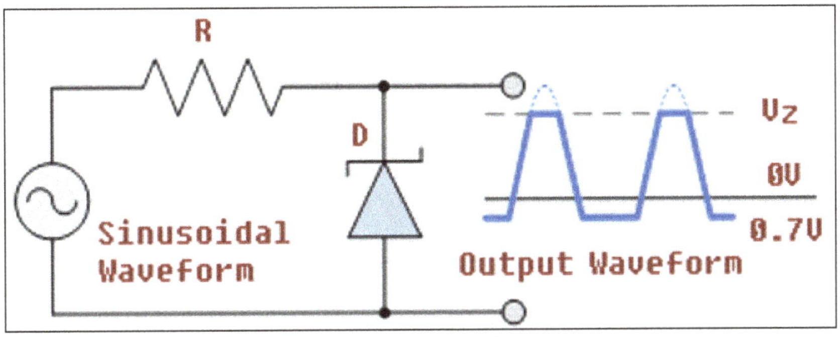

Diode Clipper Circuit.

Use of Diode for Clamping

An electronic circuit that prevents the signal from exceeding a certain defined value is known as a clamping circuit. Diode's clamping circuits are used as voltage multipliers and to remove distortions in the signal.

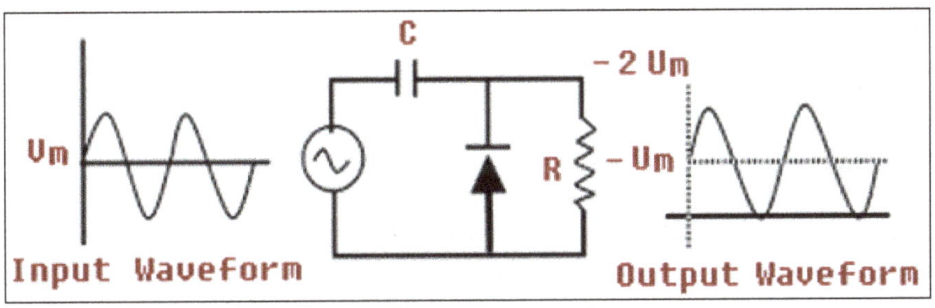

Diode Clamping Circuit.

Advantages of Diode

The advantages of Diodes include:

- Diodes are compact in size and compatible.
- Designing electronic circuit is simple with certain diodes like Zener diodes.
- Diodes help in controlling the current flow.
- These diodes produce less unwanted noise.
- Certain diodes like Schottky diodes can operate at high frequencies.
- Light emitting diodes are highly efficient compared to other diodes and they can emit light of the expected colors.
- Operates at high switching speed.

Disadvantages of Diodes

The disadvantages of Diodes include:

- Power dissipation is more for Zener diodes and hence it is less efficient for heavy loads.
- Diodes are highly sensitive to temperature.
- Amplification is necessary in photodiode based circuits.
- Light emitting diodes are expensive compared to other diodes.
- Certain diodes like Schottky diodes have low maximum reverse voltage.
- They have high reverse current and impedance.

Thyristor

Thyristor is a generic term for a bipolar semiconductor device which comprises four semiconductor layers and operates as a switch having a latched on-state and a stable off-state. Thyristors have three states:

- Reverse blocking state,
- Forward blocking state,
- Forward conducting state.

Thyristors are manufactured by diffusion. Thyristors can be turned on by applying gate signal. The thyristor has been triggered into conduction and will remain conducting until the forward current drops below a threshold value known as the holding current. Thyristors can be classified as standard or slow phase-control-type and fast-switching or inverter-type. A short duration gate pulse is sufficient to turn on the thyristor. The device with only turn on capability is referred to as —conventional thyristor, or —thyristor. various gate structures are used to manufacture thyristors in order to control the di/dt, turn-on time, and turn-off time. There are several versions of thyristors with turn-off capability.

Classification of Thyristors

Depending on the physical construction, nature of i-v characteristics and turn-on and turn-off behavior, thyristors can be classified. The different types of thyristors are:

Thyristors with Turn on capability (Unidirectional control)	Thyristors with Turn off capability (Unidirectional control)	Bidirectional Control
Phase-controlled thyristors (or SCRs).	Gate turn–off thyristors (GTOs).	Bidirectional triode thyristors (TRIACs).
Amplifying Gate thyristors (or SCRs).	MOS turn–off thyristors (MTO).	

Fast switching thyristors (or SCRs).	Emitter turn-off thyristors (ETOs).	
Asymmetrical thyristors (ASCRs).	Integrated gate commutated thyristors (IGCTs).	Bidirectional phase controlled thyristors (BCTs).
Light activated silicon controlled rectifiers (LASCRs).	MOS controlled thyristors (MCTs).	
FET–controlled thyristors (FET-CTHs).	Static induction thyristors (SITHs).	
Reverse-conducting thyristors (RCTs).		

Among the above mentioned thyristors, some are conventional and some are having turn off capability. TRIAC and BCT can conduct in both directions with gate control.

Unidirectional Thyristors with Turn-on Capability

Conventional thyristors are widely used and have only turn on capability. Those are used in line commutated converters (ac-dc, ac-ac, cycloconverters) as well as in dc choppers and inverters. For choppers and inverters, the main requirement is fast turn on and turns off. Unidirectional thyristors with turn-on capability are:

- Phase controlled thyristors (or SCRs),
- Amplifying gate thyristors (or SCRs),
- Fast switching thyristors (or SCRs),
- Asymmetrical thyristors (ASCRs),
- Light activated silicon-controlled rectifiers (LASCRs),
- FET–controlled thyristors (FET-CTHs),
- Reverse conducting thyristors (RCTs).

Phase Controlled Thyristors (PCT's or SCRs)

PCT's generally operate at the line frequency. Natural communication is used to turn off. When a gate trigger current pulse is applied to gate-cathode, a thyristor starts conduction in a forward direction and rapidly latches into full conduction with a low forward voltage drop. When the anode current comes to zero, thyristor stop conducting in a few tens of microseconds and blocks the reverse voltage. The turn-off time tq is of the order of 50 to 100μs. PCTs are most suited for low speed switching applications and hence also known as a converter thyristors.

The modern thyristors use an amplifying gate, in which an auxiliary thyristor T_A is used with the main thyristor T_M as shown in Figure below. External trigger is applied to gate of T_A turning it on. The amplified output of T_A is applied as a gate signal to the main thyristor T_M. The amplifying gate permits high dynamic characteristics with typical dv/dt of 1000 V/μs and di/dt of 500 A/μs. It reduces the values of di/dt limiting inductor and dv/dt capacitor.

The on-state voltage V_T varies typically from about 1.15V – 2.5V depending on the current. The Thyristors are available up to 5-6 kV and maximum current 4-6 kV. Because of their low cost, high efficiency, ruggedness, and high voltage and current capability, these thyristors are extensively used in line commutated converters. They are used for almost all high-voltage dc (HVDC) transmission and high voltage DC drives and supplies.

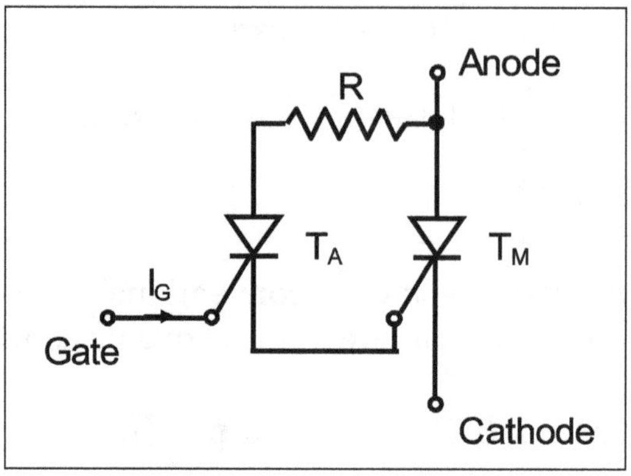

Amplifying gate Thyristor.

Features of PCTs

- Positive feedback — a latching device,
- A minority carrier device,
- Double injection leads to very low on-resistance, hence low forward voltage drops in very high,
- Voltage devices,
- Cannot be actively turned off by gate control,
- A voltage-bidirectional two-quadrant switch,
- 5kV- 6kV, 1kA – 2 kA devices.

Applications

- Line commutated converters,
- DC motors drives,
- AC/DC static switches,
- SVC – static var compensator.

Fast Switching Thyristors (or SCRs)

Turn-off time of these thyristors is small, generally in the range 5 to 50 μs. Turn off time depends on the voltage range. These are used in the high-speed switching applications with forced commutations, for example; DC choppers, forced commutated inverters and resonant inverters. Therefore, these thyristors are also known as an inverter thyristors. The on-state forward drop varies approximately as an inverse function of the turn-off time t_q.

These thyristors have high dv/dt of typically 1000 V/μs and di/dt of 1000 A/μs. The fast turn-off and high di/dt reduces the size and weight of commutating or reactive circuit components. The on-state voltage of an 1800-V, 2200-A thyristor is typically 1.7 V.

Applications

- DC –DC converters for small power drives,
- Converters for Resistive welding,
- Forced commutated inverters,
- Induction heating.

Asymmetrical Thyristor/Asymmetric Silicon Controlled Rectifier (ASCR)

ASCR is a modified version of thyristor. It is fast switching thyristor with a very limited reverse blocking capability, typically 10 V. A cross-sectional view and v-i characteristics of the ASCR is shown in Figure below. The reverse blocking capacity is reduced in ASCR by making middle 'n' layer thinner than that of SCR. The middle 'n' layer consists of low resistivity region (n^+) and high resistivity region (n^-) as shown in structure. Turn off time of ASCR is much shorter than SCR, typically 3 to 5 μs. Due to fast switching; ASCR is suitable in inverters and hence are also called as inverter thyristor.

During the reverse recovery transient the flow of reverse current causes holes to be injected across the junction J_2 from the p_2 region to the n1 region. These holes have to disappear, mainly by recombination, before the junction J_2, which is the junction responsible for blocking forward voltages, recovers its mocking ability. In normal thyristors, this recombination process takes a longer time because of the high purity level of the n_1 region. In the asymmetrical thyristor, the presence of the higher impurity n^+ region speeds up the recombination process and thus shortens the turn off time.

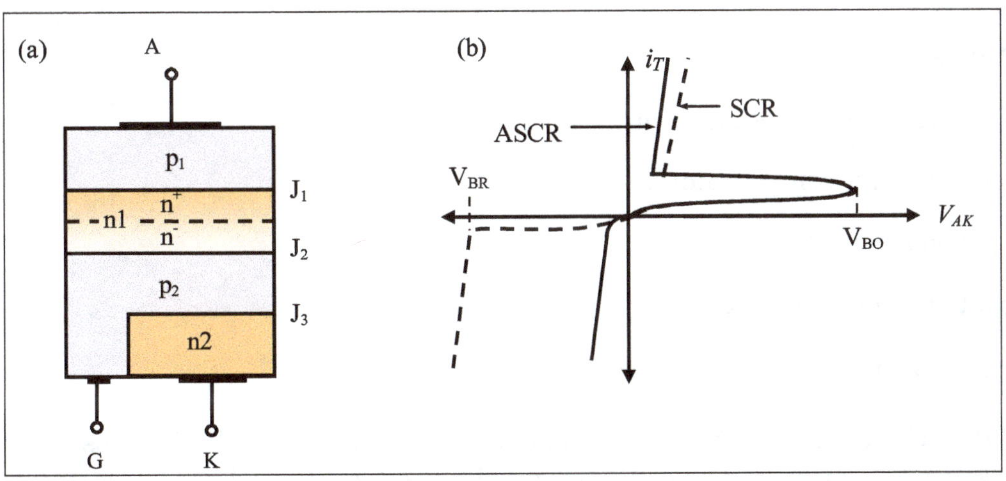

ASCR (a) Structure (b) I-Characteristics.

Applications

- Frequency converters,

- Induction heating,
- Resistive welding, electrical heating,
- DC motors control,
- Forced commutated inverters,
- Asynchronous drives,
- Battery Charging equipment.

Light Activated Silicon Controlled Rectifiers (LASCRs)

It is also called as light triggered thyristors (LTT). In LASCR, the light particles (photon) are made to strike the reverse biased junction, which causes an increase in the number of electron-hole pairs triggering the thyristor. For light triggered thyristors, a slot is made in the inner p layer. If the intensity of the light is greater than certain critical value, the thyristor will turn on.

The gate terminal is also provided externally. For practical application the resistor is connected between gate and cathode to reduce the sensitivity of gate. It increases the dv/dt capability. An LASCR offers complete electrical isolation between the light-triggering source and the switching device of a power converter, which floats at a potential of as high as a few hundred kilovolts. Due to electrical isolation capability, LASCRs are used in high-voltage and high current applications, for example, HVDC transmission and static reactive power or VAR compensation.

The voltage rating of an LASCR could be as high as 4 kV at 1500A with light triggering power of less than 100 mW. Because of this low turn-on energy, multiple cascaded amplifying gates are laterally integrated to achieve modest initial current rises limited to 300A/μs. The typical di/dt is 250 A/μs and the dv/dt could be as high as 2000 V/μs.

Applications

- HVDC transmission equipment,
- Reactive power compensators,
- High voltage drives,
- High power pulse generators.

FET–Controlled Thyristors (FET-CTHs)

In FET-CTH device contains inbuilt an n-channel enhancement MOSFET across thyristor in as shown in Figure. It is turned on by applying sufficient voltage, typically 3 V, across gate cathode. Triggering current of main thyristor is generated internally. Since it is voltage controlled device drive requirement is less that of SCR. This thyristor cannot be turned off by gate control. It has a high switching speed, high di/dt, and high dV/dt. Because of variety of turn off thyristors developed in two decades, this device could not become popular.

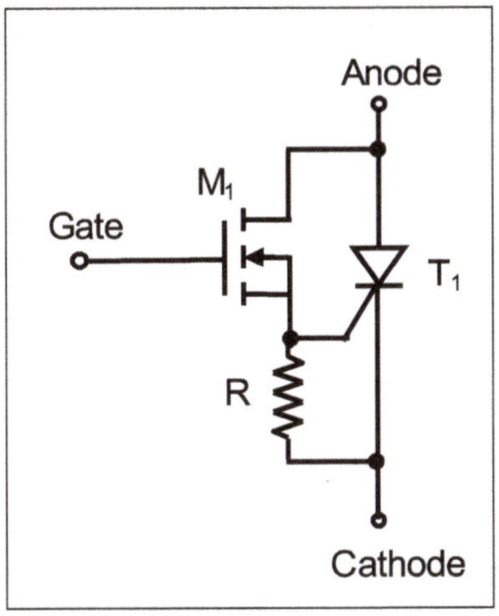

Equivalent circuit of FET-Controlled Thyristor.

Reverse Conducting Thyristors (RCTs)

In many converters and inverter circuits, an anti-parallel diode is connected across an SCR to allow a reverse current flow due to inductive load and to improve the turn-off requirement of commutation circuit. The diode clamps the reverse blocking voltage of the SCR to 1 or 2 V under steadystate conditions. However, under transient conditions, the reverse voltage may rise to 30V due to induced voltage in the circuit stray inductance within the device.

An RCT is a compromise between the device characteristics and circuit requirement as mentioned above. It has asymmetric punch through (PT) structure with an integrated anti-parallel diode. The reason for integrating the SCR and diode is to minimize the interconnecting lead inductance. The circuit symbol, and cross sectional wafer view, are shown in Figure.

Since no reverse voltage will be applied to the RCT there is only the cathode side deep p-diffused layer. This and the ASCR n-region type field stopper result in low forward voltage characteristics. As in the ASCR case, the highly n-type doped anode end of the wide n-region also allows higher forward voltages to be blocked. An RCT is also called as ASCR. The forward blocking voltage varies from 400 to 2000 V and the current rating goes up to 500 A. The reverse blocking voltage is typically 30 to 40V.

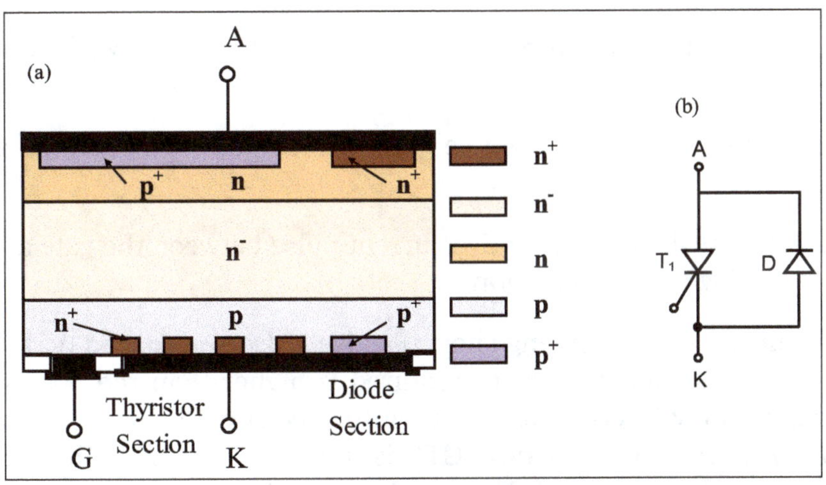

Reverse Conducting Thyristor (a) Structure (b) Equivalent circuit.

Advantages

- Compactness of a converter is obtained due to inbuilt diode.
- Undesired loop inductance effect gets eliminated.
- Unwanted reverse voltage transients gets reduced which results in the better commutation.

Applications

- DC drives for traction applications.
- High power choppers and inverters.

Unidirectional Thyristors with Turn-off Capability

The disadvantage of the conventional thyristor is no turn off capability. Hence, forced commutation circuitry is required if used in inverters, choppers. Forced commutation circuitry is bulky and heavy. Commutating chokes also produces the acoustic noise. To avoid this, thyristors with turn off capability are preferred. Many thyristors with turn off capabilty are developed in last two decades. Some of those which are used widely for different applications are:

- Gate turn–off thyristors (GTOs),
- MOS turn–off thyristors (MTO),
- Emitter turn-of (control) thyristors (ETOs),
- Integrated gate commutated thyristors (IGCTs),
- MOS–controlled thyristors (MCTs),
- Static induction thyristors (SITHs).

Gate Turn-off Thyristors (GTOs)

GTO is like an SCR, but it is fully controllable switch with turn on as well as turn-off capability using gate signal.

- Turn on is accomplished by a positive gate current pulse between the gate and cathode terminals.
- Turn off is accomplished by a negative current pulse between the gate and cathode terminals reverse biasing the gate junction.

Practically, GTO is turned on by applying short positive pulse and turned off by a short negative pulse to its gate. The negative gate current required is higher than the positive current. GTOs are used at very high power levels, and they require special gate control circuitry. A GTO is a non-latching device. A static characteristic of GTO is similar to the conventional SCR. The symbol, basic structure and equivalent circuit of GTO are shown in Figure.

Compared to SCR, there is an additional n^+ -layer near the anode that forms a turn-off circuit between the gate and the cathode in parallel with the turn-on gate. The equivalent circuit that is shown in Figure is similar to SCR, except for its internal turn off mechanism.

Turn-on: The GTO has a highly inter digited gate structure with no regenerative gate. As a consequence, a large initial gate trigger pulse is required to turn on. The turn on stages after applying gate pulse is shown in figure.

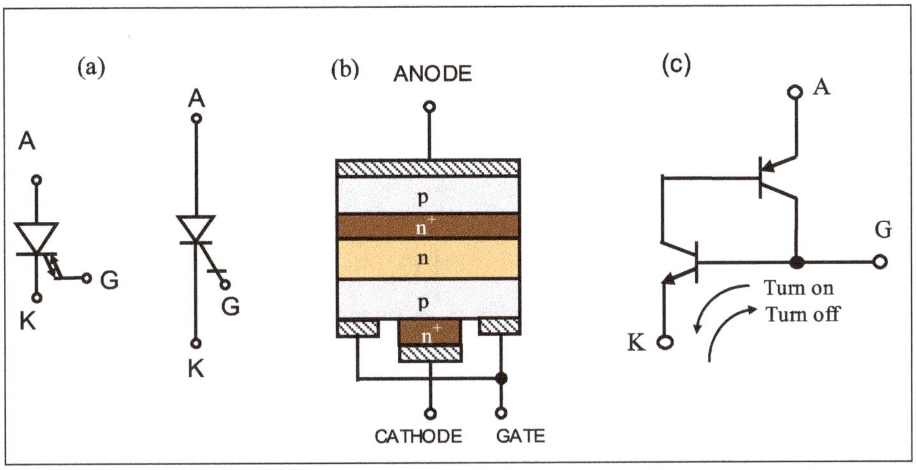

Gate turn-off thyristor (GTO) (a) symbols (b) structure (b) equivalent circuit.

If a turn on current pulse is applied to gate as shown in Figure, npn transistor Q_1 turns on which turns on the base-emitter junction of the pnp transistor Q_2. Because of that Q_2 turns on which further drives Q_1 and thyristor conducts.

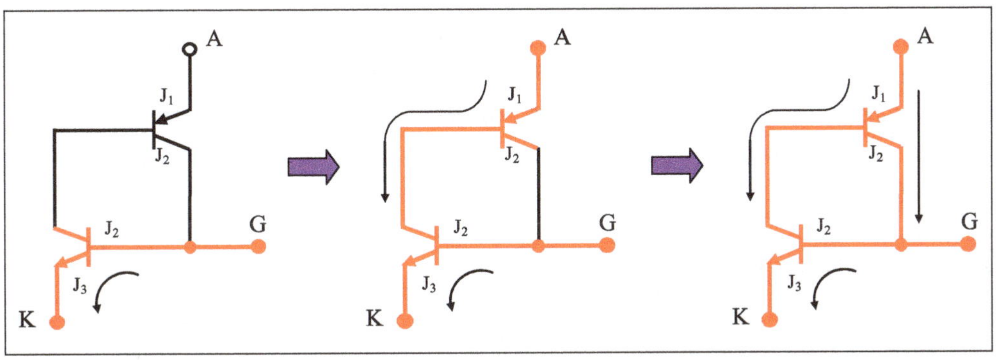

Turn on sequence of GTO.

On-state: Once the GTO is turned on, forward gate current must be continued further, to insure the device remains in conduction. Otherwise, the device cannot remains in conduction during the on-state period. After turning on the GTO, the current required to keep GTO in conduction is much less than the initial pulse magnitude, generally > 1% of the peak gate current. It ensures that the gate does not unlatch. This is illustrated with the help of turn on characteristics as shown in Figure. A typical turn on gate pulse and its effect on thyristor current and voltage are depicted in the timing diagram. Minimum and maximum values of IGM and dig/dt are given in the device data sheet. The rate of rise of gate current di_g/dt affects the device turn-on losses. A longer period is required if the anode current di/dt is low such that IGM is maintained until a sufficient level of anode current is established.

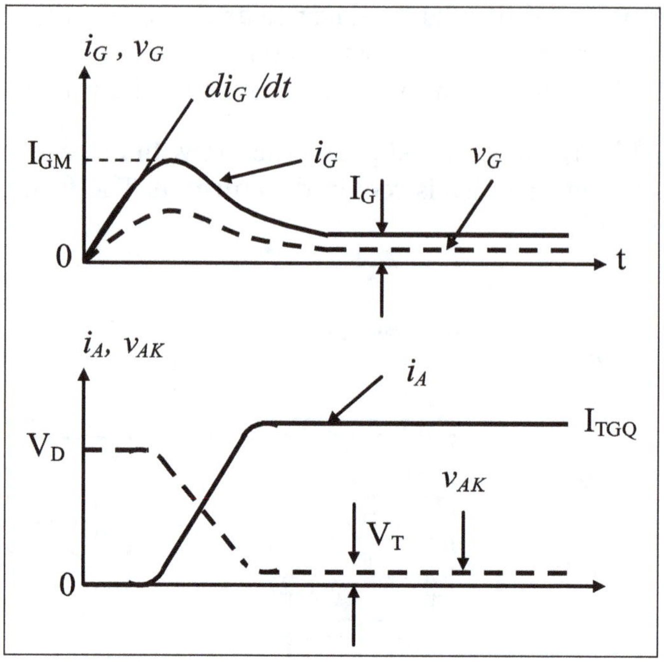

Typical turn on characteristics of GTO.

Turn off: If a large pulse current is passed from the anode to the gate, it takes away sufficient charge carriers from the cathode, that is, from the npn transistor Q_1. Thus, pnp transistor Q_2, can be drawn out of the generative action. As transistor Q_1 turns off, transistor Q_2 is left with an open base, and the GTO returns to the non-conducting state. The turn off process is illustrated by turn off sequence as shown in Figure.

The turn off performance of a GTO is greatly influenced by the characteristics of the gate turn-off circuit. Thus, the characteristics of the turn off circuit must match the device requirement. A GTO has low gain during turn-off, typically 4-6. Turn off gain is requires a relatively high negative current pulse to turn off. Turn off gain is given as,

$$\beta_m = \frac{I_A}{I_G} = \frac{\alpha_{npn}}{\alpha_{npn} + \alpha_{pnp} - 1}$$

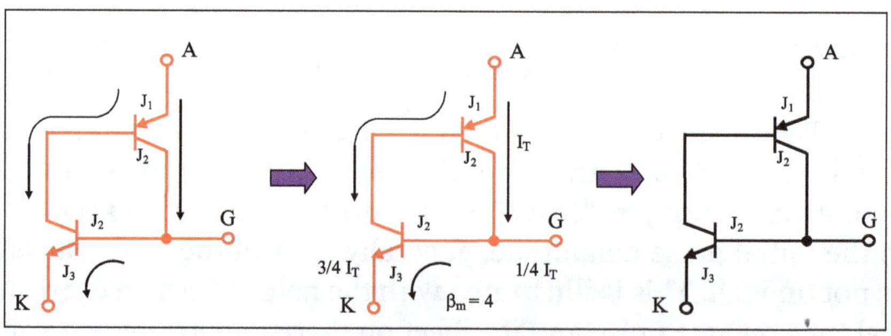

Turn off sequence of GTO.

The turn-off process involves the extraction of the gate charge, the gate avalanche period, and the anode current decay. The amount of the charge extraction is a device parameter and its value is not significantly affected by the external circuit conditions. The initial peak turn-off current and turn-off time, which are important parameters of the turning-off process, depends on the external circuit components. A typical anode current versus the turn-off pulse is shown in Figure. The GTO has a long turn-off, tail-off current at the end of the turn-off which limits the high frequency operation. The next turn-on must wait until the residual charge on the anode side is dissipated through the recombination process.

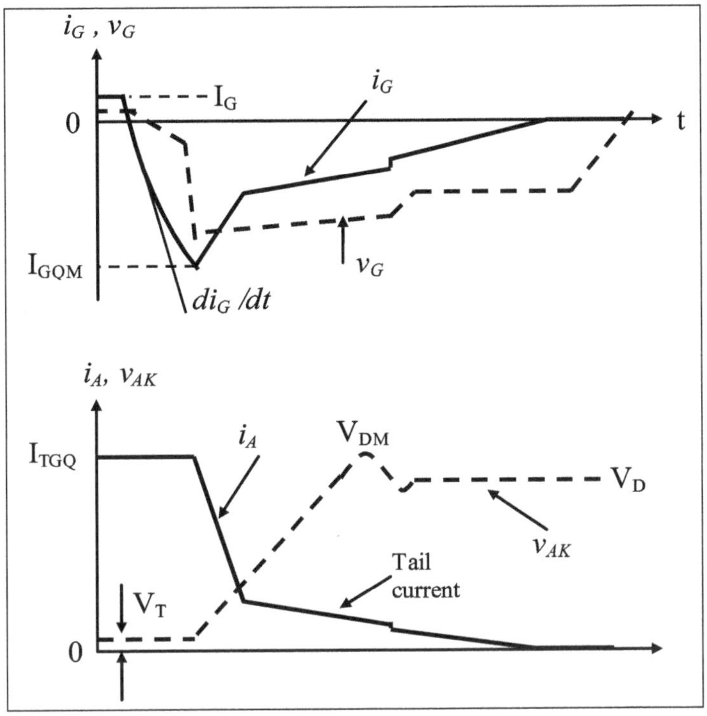

Typical turn off characteristics of GTO.

The GTO gate drive has to fulfill the four following functions:

- Turn the GTO on by means of a high current pulse (I_{GM}).
- Maintain conduction through provision of a continuous gate pulse during the on-state (IG also known as the —back-porch current).
- Turn the GTO off with a high negative gate current pulse (I_{GQM}).
- Reinforce the blocking capability of the off-state device, by negative gate voltage or, at least, by a low impedance resistor.

There are different approaches to GTO gate drive design. Figure depicts a simple gate drive circuit with turn on and turn off capability. It is suitable for both industrial and traction applications. To turn on the GTO, positive pulse to the gate transistor of T_1 turns on the transistor. The high current pulse (with peak current I_{GM}) gets applied to gate of GTO via C_1, R_1, R_2, C_2 and T_1. After some time C_2 gets charged and low current (I_G) gets applied through high resistor R_1. Thus, R_1 determines amplitude of the continuous gate current, whereas R_2 and C_2 shape the initial gate pulse (I_{GM}). Transistor T_2 and C_3 constitute the turn-off channel, and provide negative gate voltage during the GTO's blocking period. Resistor R_G guarantees minimum blocking capability for the GTO in case the gate unit power supply fails. Transistor T_2 consists of several transistors in parallel, depending on the required peak negative gate current I_{GQM}.

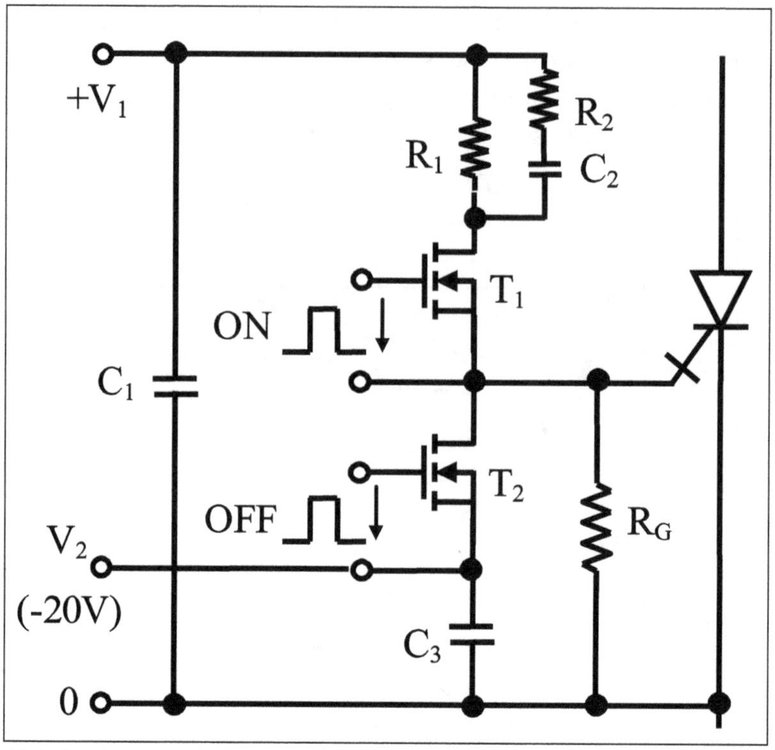

Circuit Diagram of GTO gate driver with turn on and turn off capability.

Comparison of GTOs over SCRs

- Elimination of commutating components (chokes and capacitors) in forced commutation, results in reduction in cost, weight and volume.
- Due to the elimination of commutation chokes, acoustic and electromagnetic noise is reduced.
- Faster turn-off, permits high-switching frequencies and improves efficiency of converters.
- It has higher on-state voltage than that of SCRs.

Advantages

- High current–voltage capability.
- Low conduction loss, but higher than SCR.
- Low cost due to turn off capability which eliminates forced commutation circuitry.

Disadvantages

- Non-uniform turn-off – poor RBSOA and dv/dt snubber required.
- Non-uniform turn-on – di/dt snubber required.
- Current control – high gating power.

- Long switching time due to turn off tail, hence high switching loss —long storage time, minimum ontime and off-time requirements.
- No current limitation capability – limits FBSOA.

In voltage-source converters, a fast recovery anti-parallel diode is required across each GTO. In such cases, asymmetric GTOs are used. This is achieved by introducing a heavily doped n^+-layer (buffer layer) at the end of the n-layer. Asymmetric GTOs have lower on state voltage drop and higher voltage and current ratings.

Applications

- Motor drives.
- Static VAR compensators (SVCs).
- AC/DC power supplies with high power ratings.
- Force-commutated voltage-fed thyristor inverters.

MOS Turn-off Thyristors (MTOs)

The MTO was developed by Silicon Power Company (SPCO). Figure shows the symbol, structure, and equivalent circuit of the MTO. It has two control terminals turn on gate and turn off gate. It is a combination of a GTO and an MOSFET, which overcome the limitations of the GTO turn-off ability. GTOs require a high pulse-current drive for the low impedance gate due to low turn off gain, typically 3- 5. The gate circuit must provide high gate turn-off current whose typical peak amplitude is 20-35% of the current to be controlled i.e. anode current. This drawback is overcome using MTO, in which the signal voltage is necessary to turn MOS transistor on and off. Its structure is similar to that of a GTO. MTOs are available with high voltage up to 10 kV and high current up to 4 kA. MTOs can be used in high-power applications ranging from 1 to 20 MVA.

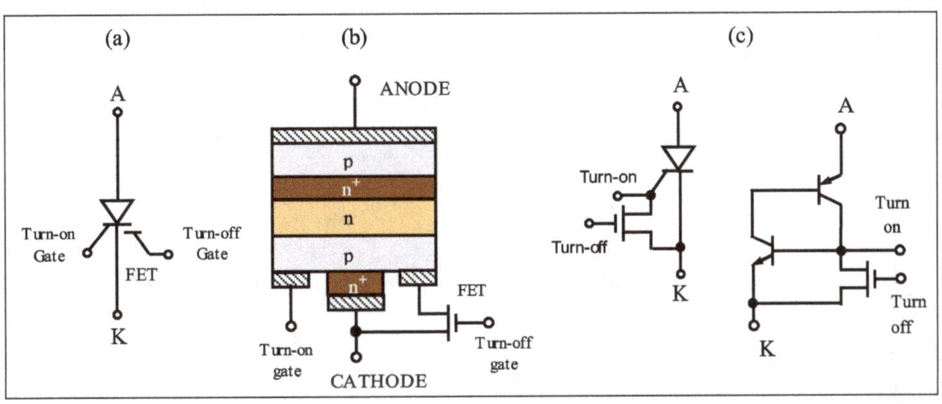

MOS Turn Off Thyristor (MTO) – (a) Symbol. (b) Structure (c) Equivalent circuits.

Turn-on: It is similar to GTO. The MTO is turned on by applying gate current pulse to the turn-on gate. Turn-on pulse turns on the npn transistor Q_1, which then turns on the pnp-transistor Q_2 latching the MTO.

Turn-off: To turn-off the MTO, a voltage pulse is applied to the MOSFET gate. Turning on the MOSFETs, shorts the emitter and base of the Q_1, thereby stopping the latching process. In contrast, a GTO is turned off by sweeping enough current out of the emitter base of npn transistor with a large negative pulse to stop the regenerative latching action. As a result, the MTO turns off much faster than a GTO (typically 1-2 μs). Therefore the losses associated with the storage time are almost eliminated. MTO has a higher dv/dt and hence requires much smaller snubber components. Similar to GTO, the MTO has a long turn-off tail of current at the end of the turn-off which limits the operating frequency of MTO.

Advantages

- High power rating (up to 10 kV and 4 kA).
- Fast switching speed (up to 5 kHz).
- Higher dv/dt and hence requires much smaller snubber components.
- Cost and gate drive power requirement is low as compared with other turn off devices.

Applications

- High voltage applications upto 20 MVA.
- Voltage source inverters for high power.
- Flexible AC line Transmissions (FACTs).
- Motor drives.

Emitter Turn-off Thyristors (ETOs)

The ETOs is MOS-GTO hybrid device that combines the advantages of both the GTO and the MOS-FET. ETO was invented at Virgina Power Electronics Center in collaboration with SPCO. Figure shows the ETO symbol, its equivalent circuit, and the pnpn structure. ETO has two gates, turn on gate and turn off gate. Turn-on gate is similar to that of GTO. Turn-off gate is the gate of MOSFET in series with GTO structure. High power ETOs with a current rating of up to 4 kA and a voltage rating of up to 6 kV are available.

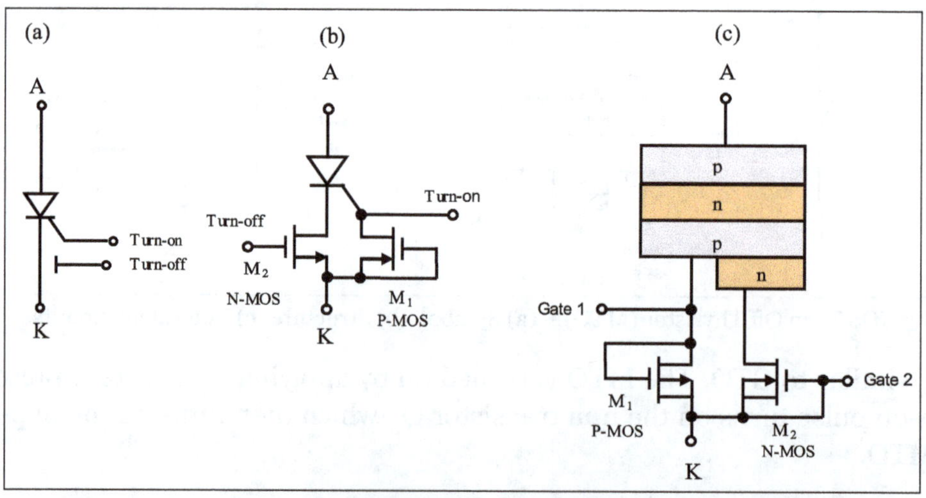

Emitter turn-off thyristor (a) Symbol, (b) Equivalent circuit (c) Structure.

Turn On: An ETO is turned on by applying positive voltages gate1 and gate2. A positive voltage to gate 2 turns on NMOS and turns off PMOS. An injection current into the GTO gate (through Gate1) turns on the ETO.

Turn off: When a turn-off negative voltage signal is applied to the gate 2 i.e. gate of NMOS M_2, it turns off and transfers all the current away from the cathode (n emitter of the npn-transistor of the GTO) into the base via gate of PMOS M_1. This stops the regenerative latching process and results in a fast turn-off. It is important to note that both MOSFETs are not subjected to high-voltage stress, no matter how high the voltage is on the ETO. This is due to the internal structure of the GTO's gate-cathode is a PN-junction. Series MOSFET, M_2 has to carry the main GTO current which increases the total voltage drop by about 0.3 to 0.5 V and corresponding power dissipation. Similar to a GTO, the ETO has a long turn-off tail of current at the end of the turn-off which limits the high frequency operation.

Advantages

- High-power rating (up to 4 kA and 6 kV).
- Fast switching speed (up to 5 kHz).
- Cost and gate drive power requirement is low as compared with other turn off devices.
- A wide reverse biased safe operation area (RBSOA).
- Snubberless turn-off capability.
- Simplicity in over-current protection.
- Capable of parallel and series operation.

Applications

- Voltage source inverters for high power.
- Flexible AC line Transmissions (FACTs).
- Motor drives.
- Static Synchronous Compensator (STATCOM).

Integrated Gate-Commutated Thyristors (IGCTs)

The integrated gate-commutated thyristor (IGCT) was introduced by ABB in 1997. The IGCT integrates a gate commutated thyristor (GCT) with a multilayered printed circuit board gate drive. Basically, it is a high-voltage, high-power, hard-driven, asymmetric blocking GTO with unity turn-off current gain. This means that a 4500V IGCT with a controllable anode current of 3000A requires turnoff negative gate current of 3000A. A very fast and large gate current pulse of full rated current draws out all the current from the cathode into the gate in about 1μs to ensure a fast turn-off.

The cross section of an IGCT and equivalent circuit of a GCT are similar to that of a GTO shown in Figure. Actually IGCT is the close integration of GTO and the gate drive circuit with multiple

MOSFETs in parallel providing the high gate currents. An IGCT may also have an integrated reverse diode, as shown by the n^{+} n^{-} p junction on the right side of Figure. Similar to a GTO, an MTO and an ETO, the n-buffer layer evens out the voltage stress across the n^{-} -layer, decreases the on-state conduction loss, and makes the devices asymmetric. The anode p-layer is made thin and lightly doped to allow faster removal of charges from the anode-side during turn-off.

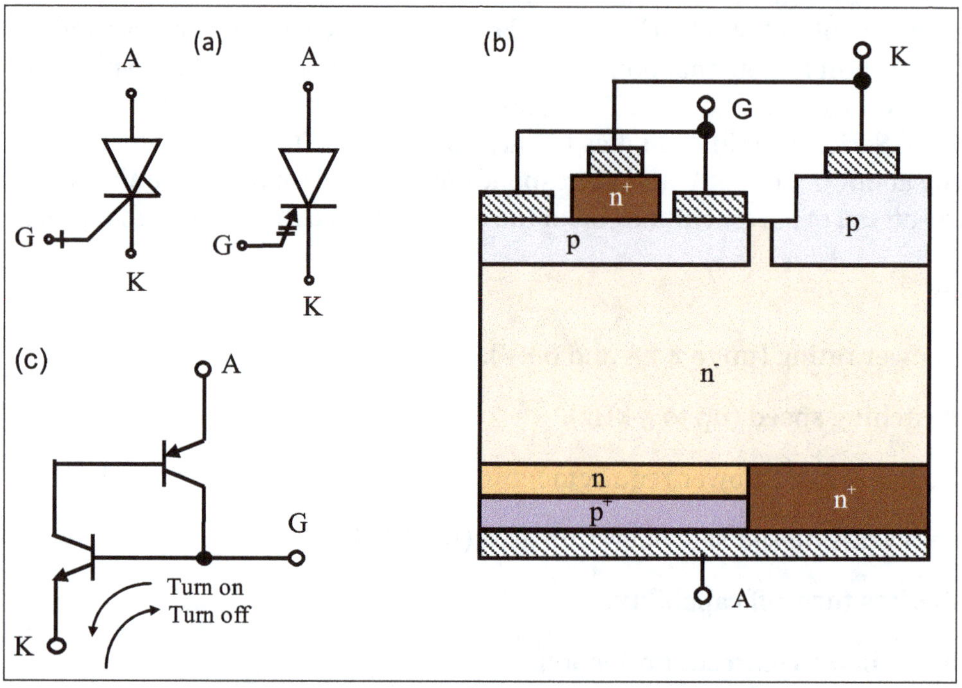

IGCT (a) Symbols, (b) Structure (c) Equivalent circuit.

Turn-on: Similar to a GTO, the IGCT is turned on by applying the turn-on current to its gate. With high gate current, turn-on is initially by npn BJT, not SCR regeneration. The pnp transistor is inoperative since the carriers in the n-base are initially ineffective since they require a finite time to transit the wide n-base.

Turn-off: IGCT is nothing but improved GTO with unity gain turn-off drive i.e. the gate current equal to the anode current. Because of that positive feedback loop of thyristor is broken and npn transistor turns off first. Therefore, thyristor turns off in open-base pnp transistor mode. The turn-off sequence from on state is depicted using Figure.

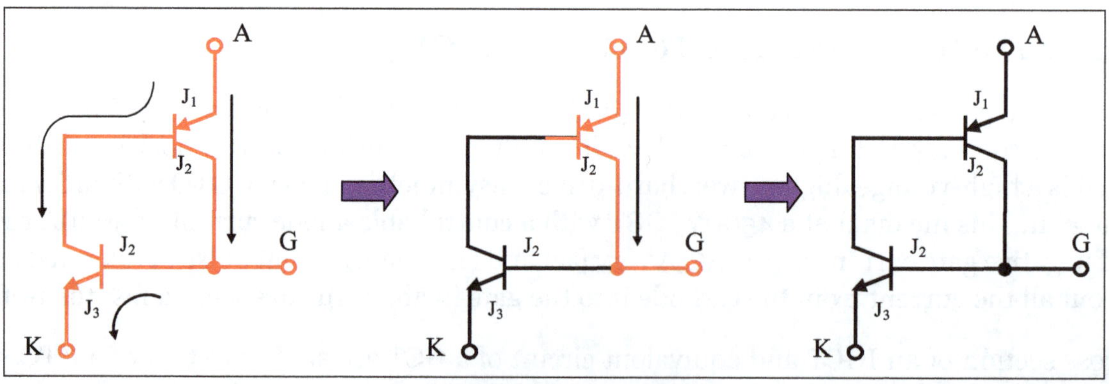

Turn off sequence of GTO.

The IGCT is turned off by a multilayered gate-driver circuit board. Driver circuit can supply a fast rising turn-off pulse, for example, a gate current of 4 kA/μs with a gate-cathode voltage of 20V only. With this rate of gate current, the npn-transistor is totally turned off within about 1 μs and the anode-side pnp transistor is effectively left with an open base and it is turned-off almost immediately. Due to a very short duration pulse, the gate-drive energy is greatly reduced and the gate drive energy consumption is minimized. The gate-drive power requirement is decreased by a factor of five compared with that of the GTO. To apply a fast-rising and high-gate current, the IGCT incorporates a special effort to reduce the inductance of the gate circuitry as low as possible. This feature is also necessary for gate-drive circuits of the MTO and ETO. The high reverse gate current results in a very short saturation delay time, enabling accurate turn-off synchronization necessary for devices to be series connected.

Gate Driver: The key to achieve a hard-driven or unity-gain turn-off condition lies in the gate current commutation rate. A rate as high as 6 kA/μs is required for 4-kA turn-off of GTO to achieve turn off. The gate drive circuit is built-in on the device module. It is necessary to hold the gate loop inductance low enough (LG ≤ 3 nH) so that a required dc gate (18 to 22 V) voltage less than the breakdown voltage of the gate–cathode junction can generate a slew rate of 6 kA/μs. High cost associated with the low-inductance housing design for the GTO. Typical GTO gate drive configuration with a small gate inductance LG is shown in Figure.

The SCR on-state regenerative mechanism is avoided at both turn-off and turn-on switching transitions thereby yielding a device more robust than the GTO. As with the GTO, an inductive series turn-on snubber is still required to cope with the initial high di/dt current. The GCT switch is thermally limited, rather than frequency limited as with the conventional GTO. Multiple IGCTs can be connected in series or in parallel for higher power applications.

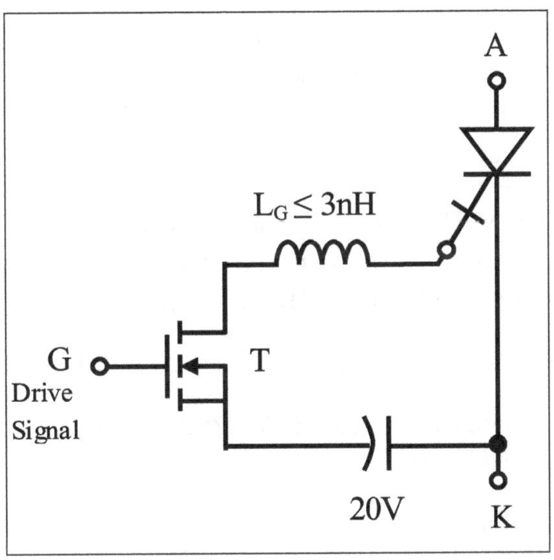

Typical GTO gate drive configuration with a small gate inductance L_G.

Advantages of IGCTs Over GTOs

- Low conduction drop.
- High Small minority carrier storage time, turn on di/dt, and turn-off dv/dt.

- Low gate driver loss.
- Faster switching of the device permits snubberless operation.
- Higher switching frequency.

Disadvantage

- High cost associated with the low-inductance housing design for the GTO.
- High cost associated with low inductance and high current design for the gate driver.

Applications

- High-power converters in excess of 100MVA.
- Static vol-ampere reactive (VAR) compensators.
- Converters for distributed generation such as wind power.

MOS Controlled Thyristors (MCTs)

An MCT is a regenerative four layer thyristor with a MOS gate structure. An MCT is an improvement over a thyristor with a pair of MOSFETs to turn on and turn off. In general, there are two types of MCT, p-channel MCT (p-MCT) and n-channel MCT (n-MCT). The p-MCT is widely used because of low on state voltage as compared with n-MCT. A symbol, structure and equivalent circuit of a p-MCT cell is shown in Figure.

The npnp structure may be represented by an npn transistor Q_1 and a pnp-transistor Q_2. The MOS-gate structure can be represented by a p-channel MOSFET (M_1) and an n-channel MOSFET (M_2). Due to an npnp structure, instead of the pnpn structure of SCR, the anode serves as the reference terminal. The gate signals are applied with reference to anode. Since gate signal of the p-MCT is applied with respect to the anode instead of the cathode, it is sometimes referred to as complementary MCT (C-MCT). In case of n-MCT the cathode serves as the reference terminal. The gate signals are applied with reference to cathode. It is turned on by a negative voltage pulse at the gate with respect to the anode and is turned off by a positive voltage pulse:

- Negative gate-anode voltage turns PMOS (M_1) on, latching both transistors Q_1 and Q_2.
- Positive gate-anode voltage turns NMOS (M_2) on, reverse biasing the base-emitter junction of Q_2 and turning off the device.
- Maximum current that can be interrupted is limited by the on-resistance of NMOS (M_2).

The device has a microcell construction. In a practical MCT, about 100,000 cells similar to the one shown in Figure are paralleled to achieve the desired current rating. Each cell contains a wide-base npn-transistor and a narrow-base pnp-transistor. Each pnp-transistor in a cell is provided with an NMOS across emitter and base to provide higher current for turn off. But, a small percentage (around 4%) of pnp-transistors is provided with PMOS across its emitter and collector to provide sufficient current to turn on.

Turn on: When a p-channel MCT is in the forward blocking state, it can be turned on by applying a negative pulse to its gate with respect to the anode. An MCT remains in the on-state until the device current is reversed or a turn-off pulse is applied to its gate.

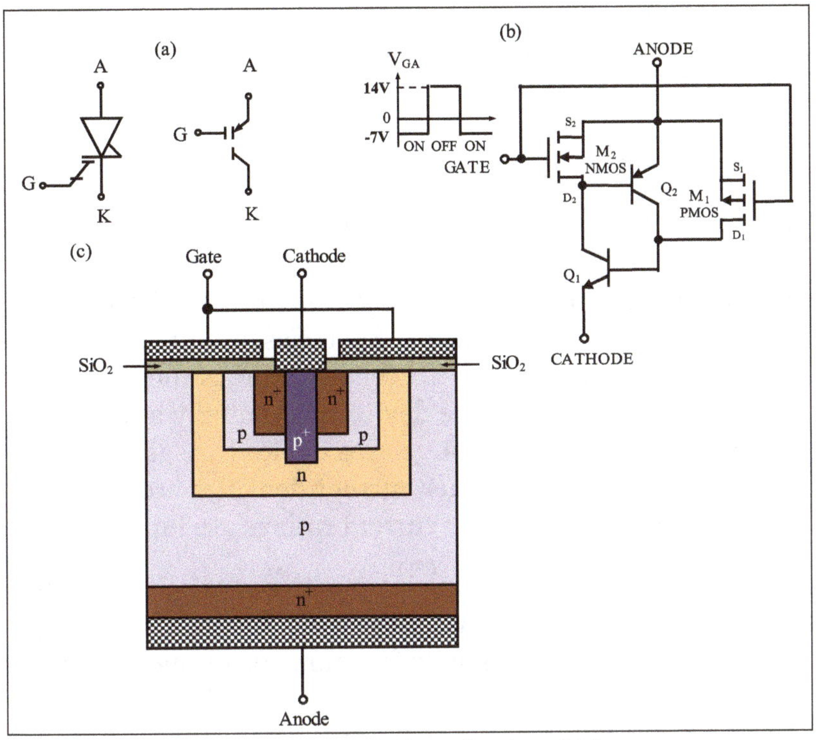

MCT (a) Symbol, (b) equivalent circuit and (c) structure.

Turn off: When a p-channel MCT is in the on-state, it can be turned off by applying a positive pulse to its gate with respect to the anode. When an n-channel MCT is in the on-state, it can be turned off by applying a negative pulse to its gate with respect to the cathode.

Attempting to turn off the MCT at currents higher than its rated peak controllable current may result in destroying the device. For higher values of current, the MCT has to be commutated off like a SCR. The gate pulse widths are not critical for smaller device currents. For larger currents, the width of the turn-off pulse should be larger. The gate draws a peak current during turn-off.

The MOS structure is spread across the entire surface of the device resulting fast turn-on and turn-off with low-switching losses. The power or energy required for the turn-on and turn-off is very small, and the delay time due to the charge storage is also very small. An MCT has:

- A low on state voltage around 1V like SCR, hence low conduction loss.
- A fast turn-on time, typically 0.4ms, and a fast turn-off time, typically 1.25ms for an MCT of 500 V, 300 A.
- Low switching losses.
- A low reverse voltage blocking capability.
- High gate input impedance, which greatly simplifies the drive circuit.

- A limited safe operating area (SOA), and therefore a snubber circuit is mandatory.
- An asymmetric voltage-blocking capability.

The device has a limited safe operating area; therefore, a snubber circuit is mandatory in an MCT converter. Also, it has complex geometry. These disadvantages have hampered its application, and the MCT has not gained widespread acceptance in the power electronics community.

Static Induction Thyristors (SITHs)

The SITH is also known as field-controlled diode (FCD) or field controlled thyristor (FCTh). It contains containing a gate structure that can shut down anode current flow. This device was first introduced by Teszner in the 1960s. It is minority carrier device, a JFET structure with an additional injecting layer. Since it is a minority carrier device, SITH has low on-state resistance and therefore low voltage drop. The cross section of a half SITH cell structure is shown in Figure. Its symbol and equivalent circuit is also shown in Figure. The device is essentially a pin diode with a gate structure that can pinch-off anode current flow. Large area devices are generally the buried-gate type because larger cathode areas and, hence, larger current densities are possible. SITH devices can have high voltage ratings up to 2.5 kV, but low current ratings are limited to 500 A.

Turn-on: A SITH is normally turned on by applying positive gate voltage with respect to the cathode. Providing sufficient positive gate current and voltage, the gate cathode p-i-n diode turns on and injects electrons into the channel (anode drift region) resulting in large conductivity modulation. A portion of whole current flows through the p^+ gate and the channel toward the cathode directly. The remaining hole current flows through the p^+ gate to the channel as the gate current of the Bipolar Mode JFET (BMFET). Consequently, there is small on state resistance resulting low on state voltage, even at large currents.

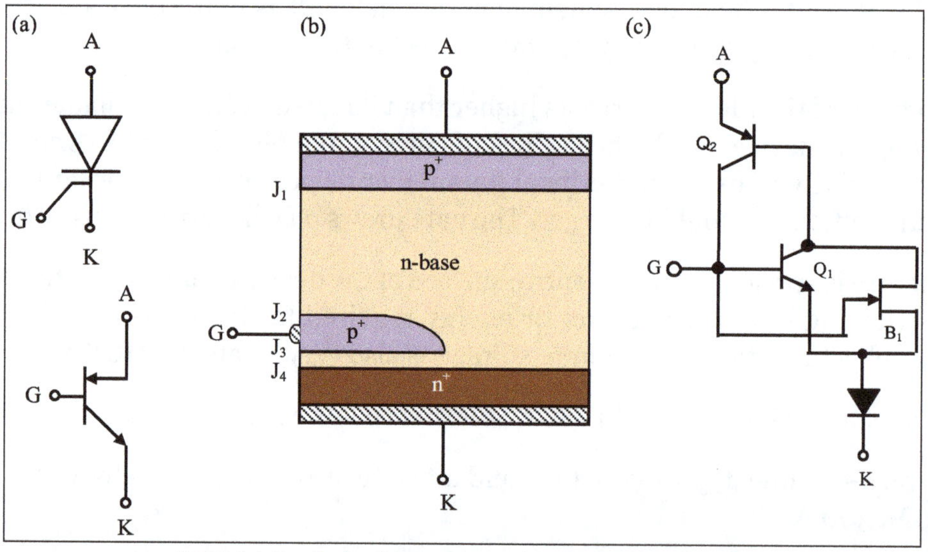

SITH (a) Symbols, (b) Structure (c) Equivalent circuit.

Turn-off: An SITH is normally turned off by applying a large reverse bias across gate and cathode. Because of large reverse bias of gate cathode junction, the depletion region of the gate junction grows and pinches off the channel connecting anode and cathode preventing current flow.

Because of addition of the pn junction at anode SITH also block high reverse voltage. The device does not have regenerative turn on and turn off i.e. it does not latch on or off. If the device is on, the removal of gate drive will cause device to turn-off.

Key Features

- It is a fast-switching device. The switching time is 1 to 6 μs.
- High dv/dt and di/dt capabilities.
- The voltage rating can go up to 2500 V.
- The current rating is limited to 300A.
- This device is highly process sensitive. Small perturbations in the manufacturing process would produce major changes in the device characteristics.

4

Laws and Theorems of Electrical Circuits

An electrical circuit is a closed loop which provides a path for the flow of electrons. There are three basic circuit elements which are necessary for the electric current to flow safely through an electrical circuit, namely, a source of energy connecting wires and devices which use curent. All the diverse aspects of electrical circuits such as the circuit diagram, phasor diagram, electrical impedance, Ohm's law, etc. have been carefully analyzed in this chapter.

Electrical Circuit

An electrical circuit is a path or line through which an electrical current flows. The path may be closed (joined at both ends), making it a loop. A closed circuit makes electrical current flow possible. It may also be an open circuit where the electron flow is cut short because the path is broken. An open circuit does not allow electrical current to flow.

Below is a basic set of symbols that you may find on circuit diagrams,

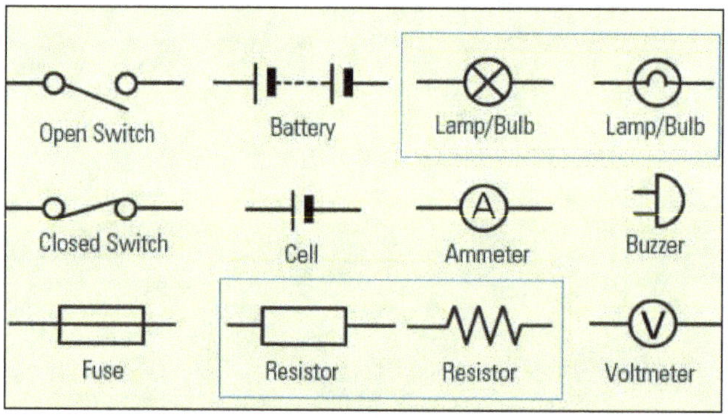

It is very important to know the basic parts of a simple circuit and the symbols that relate to them. A simple circuit has conductors, a switch, a load and a power source. Here are the functions of each part:

Conductors

These are usually copper wires with no insulation. They make the path through which the electricity flows. One piece of the wire connects the current from the power source (cell) to the load. The other piece connects the load back to the power source.

Switch

The switch is simply a small gap in the conductor where you can close or open the circuit. When the switch is closed, the circuit is closed and electricity flows.

The Load

The load is a small light bulb or buzzer that lights when the circuit is turned on. The load is also known as a resistor.

Cell

The power source is a cell. (Note that more than one cell put together is known as a battery).

The diagram below shows how a basic circuit looks like.

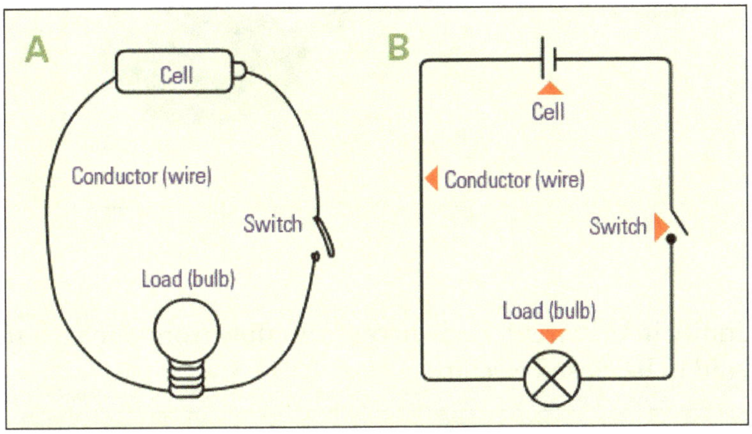

It is important to draw circuits with clean straight lines, as shown in diagram B. Avoid realistic sketches. It is important to know that a circuit can have more than the basic components in the diagram. It can have two or more batteries or two or more bulbs.

Basic Properties of Electric Circuits

- A circuit is always a closed path.
- A circuit always contains at least an energy source which acts as a source of electrons.
- The electric elements include uncontrolled and controlled source of energy, resistors, capacitors, inductors, etc.
- In an electric circuit flow of electrons takes place from negative terminal to positive terminal.
- Direction of flow of conventional current is from positive to negative terminal.
- Flow of current leads to potential drop across the various elements.

Types of Electric Circuit

The electric circuit can be categorized in three different ways:

- Open circuit
- Closed circuit
- Short circuit

Open Circuit

If due to disconnection of any part of an electric circuit if there is no flow of current through the circuit, is said to be an open circuited.

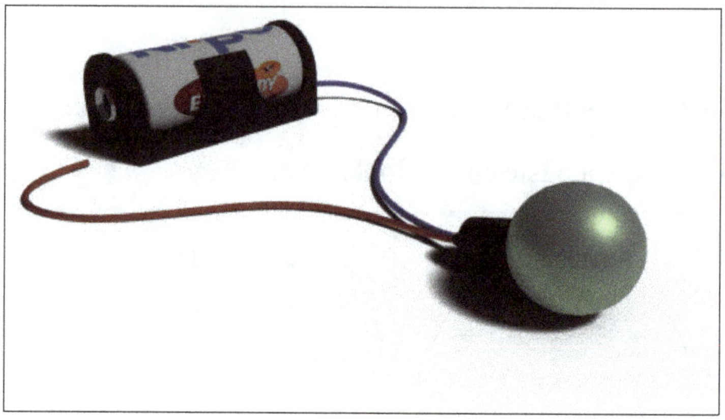

Closed Circuit

If there is no discontinuity in the circuit and current can flow from one part to another part of the circuit, the circuit is said to be closed circuit.

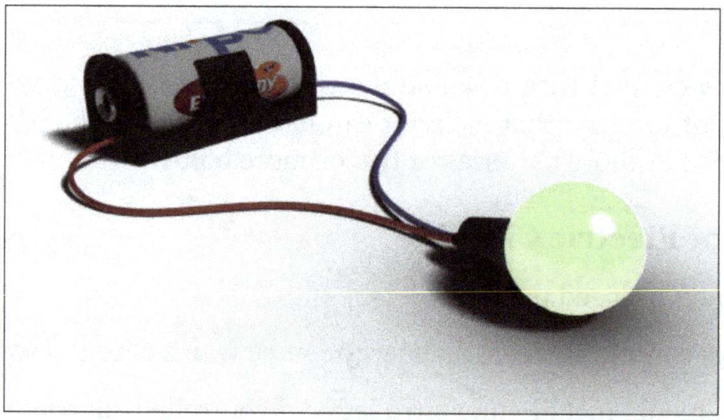

Short Circuit

If two or more phases, one or more phases and earth or neutral of AC system or positive and negative wires or positive or negative wires and earth of DC system touch together directly or connected together by a zero impedance path then the circuit is said to be short circuited.

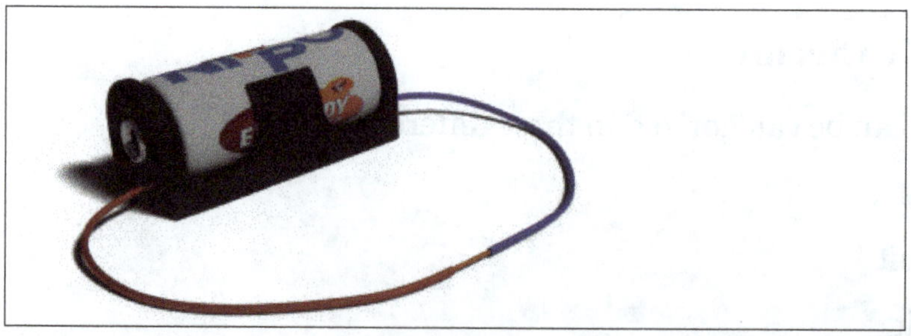

Series Circuit

A series circuit is one that has more than one resistor, but only one path through which the electricity (electrons) flows. From one end of the cell (battery), the electrons move along one path with NO branches, through the resistors, to the other end of the cell. All the components in a series circuit are connected end-to-end.

A resistor in a circuit is anything that uses some of the power from the cell. In the example below, the resistors are the bulbs. In a series circuit, the components are arranged in a line, one after the other.

Take a look at the diagram below:

The schematic drawing is a better way to draw a series circuit.

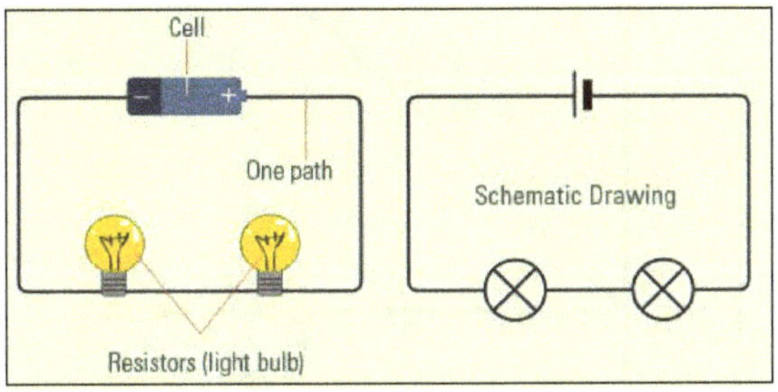

Each time there is damage (break) in any one of the resistors the entire circuit will not function. For example, if one light bulb goes out, all the other lights will go off because the electricity path in the broken bulb is cut off.

Do you put Christmas lights on the trees at home during Christmas? If the lights are in a series circuit, one burned out bulb will keep all the lights off. That is one disadvantage of series circuits. One advantage though is that you will always know if there is a break in a series circuit.

If there are many bulbs in a circuit with a battery (cell), it is very likely that the light will be dimmer because many resistors are acting on the same voltage of power from the battery.

RL Circuit

A circuit that contains a pure resistance R ohms connected in series with a coil having pure inductance of L (Henry) is known as R L Series Circuit. When an AC supply voltage V is applied the current, I flows in the circuit. IR and IL will be the current flowing in the resistor and inductor respectively, but the amount of current flowing through both the elements will be same as they are connected in series with each other. The circuit diagram of RL Series Circuit is shown below:

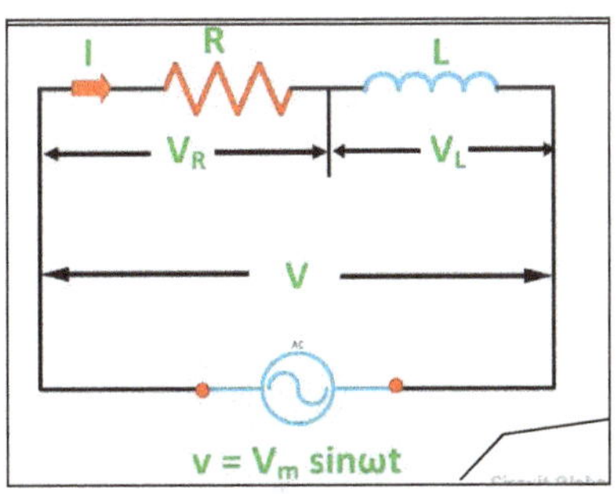

Where,

- V_R – voltage across the resistor R
- V_L – voltage across the inductor L
- V – Total voltage of the circuit

Phasor Diagram of the RL Series Circuit

The phasor diagram of the RL Series circuit is shown below:

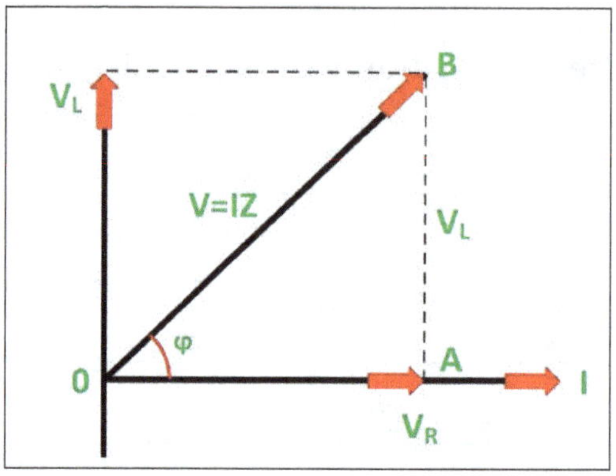

Steps to draw the Phasor Diagram of RL Series Circuit. The following steps are given below which are followed to draw the phasor diagram step by step:

- Current I is taken as a reference.
- The Voltage drop across the resistance $V_R = I_R$ is drawn in phase with the current I.
- The voltage drop across the inductive reactance $V_L = IX_L$ is drawn ahead of the current I. As the current lags voltage by an angle of 90 degrees in the pure Inductive circuit.
- The vector sum of the two voltages drops V_R and V_L is equal to the applied voltage V.

Now,

In right angle triangle OAB,

$$V_R = I_R \text{ and } V_L = IX_L \text{ where } X_L = 2\pi fL$$

$$V = \sqrt{(V_R)^2 + (V_L)^2} = \sqrt{(IR)^2 + (IX_L)^2}$$

$$V = \sqrt{R^2 \; X_L^2} \quad \text{or}$$

$$I = L = \frac{V}{Z}$$

Where,

$$Z = \sqrt{R^2 \; X_L^2}$$

Z is the total opposition offered to the flow of alternating current by an *RL* Series circuit and is called impedance of the circuit. It is measured in ohms (Ω).

Phase Angle

In *RL* Series Circuit the current lags the voltage by 90-degree angle known as phase angle. It is given by the equation,

$$\tan\varphi = \frac{V_L}{V_R} = \frac{IX_L}{IR} = \frac{X_L}{R} \quad \text{or}$$

$$\varphi = \tan^{-1}\frac{X_L}{R}$$

Power in R L Series Circuit

If the alternating voltage applied across the circuit is given by the equation,

$$v = V_m \text{Sin}\,\omega t$$

The equation of current I is given as,

$$i = I_m \sin(\omega t - \varphi)$$

Than the instantaneous power is given by the equation,

$$p = vi$$

Putting the value of v and i from the $v = V_m \text{Sin}\,\omega t$ and $i = I \ \sin(\omega t - \varphi)$ equations in the $p = vi$ equation we will get,

$$P = (V_m \text{Sin}\,\omega t) \, x \, I_m \sin(\omega t - \varphi)$$

$$p = \frac{V_m I_m}{2} 2\sin(\omega t - \varphi)\sin\omega t$$

$$P = \frac{V_m}{\sqrt{2}}\frac{I_m}{\sqrt{2}}\left[\cos\varphi - \cos(2\omega t - \varphi)\right]$$

$$P = \frac{V_m}{\sqrt{2}}\frac{I_m}{\sqrt{2}}\cos\varphi - \frac{V_m}{\sqrt{2}}\frac{I_m}{\sqrt{2}}\cos(2\omega t - \varphi)$$

The average power consumed in the circuit over one complete cycle is given by the equation shown below:

$$P = \text{average of}\,\frac{V_m}{\sqrt{2}}\frac{I_m}{\sqrt{2}}\cos\varphi - \text{average of}\,\frac{V_m}{\sqrt{2}}\frac{I_m}{\sqrt{2}}\cos(2\omega t - \varphi) \quad \textit{or}$$

$$P = \frac{V_m}{\sqrt{2}}\frac{I_m}{\sqrt{2}}\cos\varphi - \text{Zero} \qquad \textit{or}$$

$$P = V_{r.m.s} I_{r.m.s} \cos\varphi = VI\cos\varphi$$

Where, *cosϕ* is called the power factor of the circuit.

$$\cos\varphi = \frac{V_R}{V} = \frac{IR}{IZ} = \frac{R}{Z}$$

The power factor is defined as the ratio of resistance to the impedance of an AC Circuit.

Putting the value of *V* and *cosϕ* from the above equation the value of power will be,

$$P = (IZ)(I)(R/Z) = I^2R$$

From the equation above, it can be concluded that the inductor does not consume any power in the circuit.

Waveform and Power Curve of the RL Series Circuit

The waveform and power curve of the RL Series Circuit is shown below:

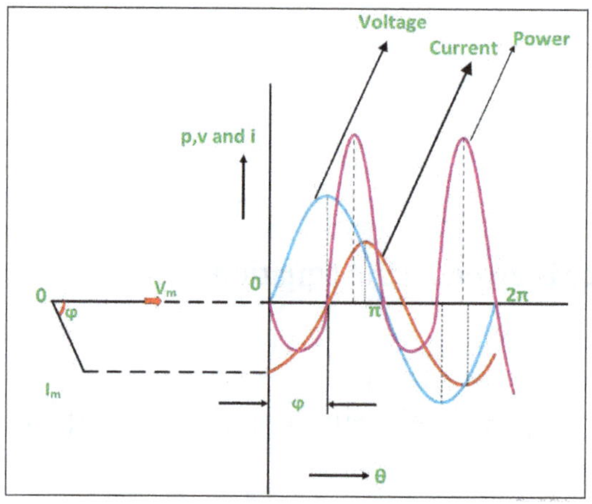

The various points on the power curve are obtained by the product of voltage and current. If you analyze the curve carefully, it is seen that the power is negative between angle 0 and ϕ and between 180 degrees and (180 + ϕ) and during the rest of the cycle the power is positive. The current lags the voltage and thus they are not in phase with each other.

Example

Calculating Characteristic Time and Current in an RL Circuit

(a) What is the characteristic time constant for a 7.50 mH inductor in series with a 3.00 Ω resistor?

(b) Find the current 5.00 ms after the switch is moved to position 2 to disconnect the battery, if it is initially 10.0 A.

Strategy for (a)

The time constant for an *RL* circuit is defined by $\tau = L/R$.

Solution for (a)

Entering known values into the expression for τ given in $\tau = L/R$ yields,

$$\tau = \frac{L}{R} = \frac{7.50mH}{3.00\Omega} = 2.50ms$$

Discussion for (a)

This is a small but definitely finite time. The coil will be very close to its full current in about ten time constants, or about 25 ms.

Strategy for (b)

We can find the current by using $I = I_0 e^{-t/T}$, or by considering the decline in steps. Since the time is twice the characteristic time, we consider the process in steps.

Solution for (b)

In the first 2.50 ms, the current declines to 0.368 of its initial value, which is

$$\begin{aligned} I &= 0.368 I_0 = (0.368)(10.0\text{ A}) \\ &= 3.68 A \; at \; t = 2.50 ms \end{aligned}$$

After another 2.50 ms, or a total of 5.00 ms, the current declines to 0.368 of the value just found. That is,

$$\begin{aligned} I' &= 0.368 I = (0.368)(3.68\text{ A}) \\ &= 1.35 A \; at \; t = 5.00 ms \end{aligned}$$

Discussion for (b)

After another 5.00 ms has passed, the current will be 0.183 A; so, although it does die out, the current certainly does not go to zero instantaneously.

RC Circuit

The combination of a pure resistance R in ohms and pure capacitance C in Farads is called RC circuit. The capacitor store energy and the resistor connect in series with the capacitor control the charging and discharging of a capacitor. The RC circuit is used in camera flashes, pacemaker, timing circuit etc.

The RC signal filters the signals by blocking some frequencies and allowing others to pass through it. It is also called first order RC circuit and is used to filter the signals by passing some frequencies and blocking others. The RC fitlers are mostly use for selecting signals and for rejecting noise.

The high-pass filter and the low-pass filter are the most common type of RC filters. The high pass filter passes the frequency greater than the fixed cutoff frequency and blocks the frequency lower than the fixed cutoff frequency. Similarly, the low-pass filter allows the frequency lower than the fixed cutoff frequency and attenuates the frequency higher than the fixed cutoff frequency.

RC Series Circuit

A circuit containing resistance and capacitance connected in series together is called an RC series circuit.

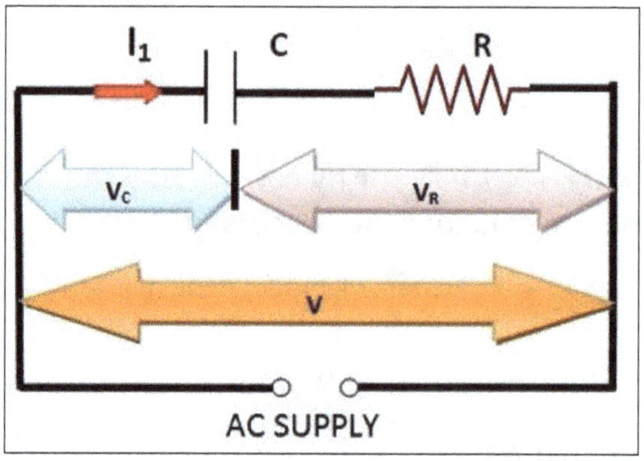

Steps to Draw a Phasor Diagram for an RC Circuit

1. Current I is taken as reference.
2. Voltage drop in resistance is (V_R):

   V_R = IR is drawn in phase with the current I.
3. Voltage drop in capacitive reactance is (V_C):

   V_C = IXC and is drawn 90° behind the current (as current leads the voltage by 90° in pure capacitive load circuit).
4. The vector sum of the two voltage drops is equal to the applied voltage (V).

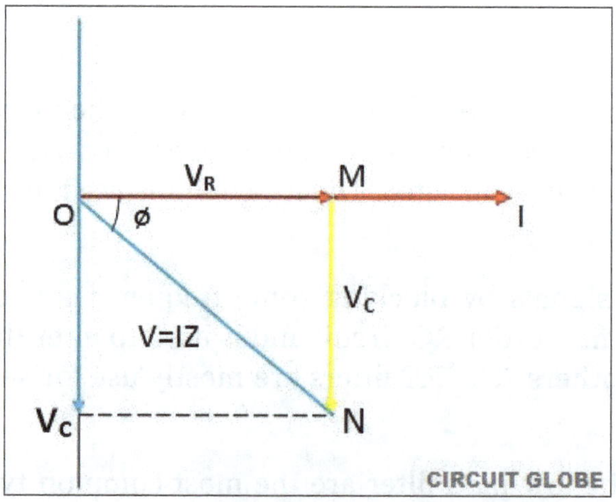

Charging

Figure below shows a simple RC circuit that employs a DC voltage source. The capacitor is initially uncharged. As soon as the switch is closed, current flows to and from the initially uncharged capacitor. As charge increases on the capacitor plates, there is increasing opposition to the flow of charge by the repulsion of like charges on each plate.

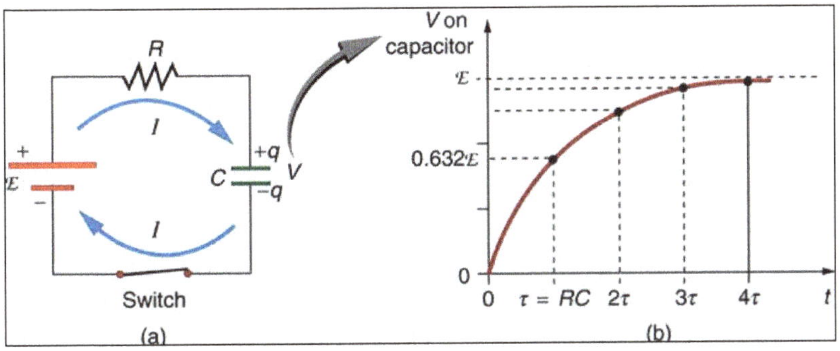

Charging an RC Circuit:

(a) An *RC* circuit with an initially uncharged capacitor. Current flows in the direction shown as soon as the switch is closed. Mutual repulsion of like charges in the capacitor progressively slows the flow as the capacitor is charged, stopping the current when the capacitor is fully charged and *Q=C·emf.*

(b) A graph of voltage across the capacitor versus time, with the switch closing at time *t=o.* (Note that in the two parts of the figure, the capital script *E* stands for emf, *q* stands for the charge stored on the capacitor, and τ is the *RC* time constant).

In terms of voltage, across the capacitor voltage is given by $V_c=Q/C$, where *Q* is the amount of charge stored on each plate and *C* is the capacitance. This voltage opposes the battery, growing from zero to the maximum emf when fully charged. Thus, the current decreases from its initial value of $I_o=emf/R$ to zero as the voltage on the capacitor reaches the same value as the emf. When there is no current, there is no IR drop, so the voltage on the capacitor must then equal the emf of the voltage source.

Initially, voltage on the capacitor is zero and rises rapidly at first since the initial current is a maximum. Figure (b) above shows a graph of capacitor voltage versus time (*t*) starting when the switch is closed at *t=o.* The voltage approaches emf asymptotically since the closer it gets to emf the less current flows. The equation for voltage versus time when charging a capacitor *C* through a resistor *R*, is:

$$V(t) = \text{emf}\left(1 - e^{t/RC}\right),$$

where, *V(t)* is the voltage across the capacitor and emf is equal to the emf of the DC voltage source. (The exact form can be derived by solving a linear differential equation describing the RC circuit, but this is slightly beyond the scope of this Atom.) Note that the unit of RC is second. We define the time constant τ for an RC circuit as τ=RC. τ shows how quickly the circuit charges or discharges.

Discharging

Discharging a capacitor through a resistor proceeds in a similar fashion, as illustrates. Initially, the current is $I_o = V_o/R$, driven by the initial voltage Vo on the capacitor. As the voltage decreases, the current and hence the rate of discharge decreases, implying another exponential formula for *V*. Using calculus, the voltage *V* on a capacitor *C* being discharged through a resistor *R* is found to be,

$$V(t) = V_0 e^{-t/RC}$$

Impedance

Impedance is the measure of the opposition that a circuit presents to the passage of a current when a voltage is applied.

Rather than solving the differential equation relating to circuits that contain resistors and capacitors, we can imagine all sources in the circuit are complex exponentials having the same frequency. This technique is useful in solving problems in which phase relationship is important. The phase of the complex impedance is the phase shift by which the current is ahead of the voltage.

Complex Analysis

For an RC circuit in, the AC source driving the circuit is given as:

$$V_{in}(t) = Ve^{j\omega t},$$

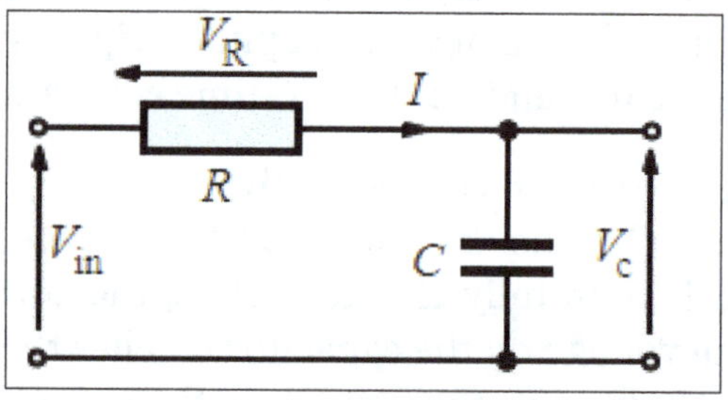

Series RC circuit.

where V is the amplitude of the AC voltage, j is the imaginary unit ($j^2=-1$), and ω is the angular frequency of the AC source. Two things to note:

1. We use lower case alphabets for voltages and sources to represent that they are alternating (i.e., we use $v_{in}(t)$ instead of $V_{in}(t)$).
2. The imaginary unit is given the symbol "j", not the usual "i". "i" is reserved for alternating currents.

Complex Impedance

The advantage of assuming sources take this form is that all voltages and currents in the circuit are also complex exponentials (having the same frequency as the source). To appreciate the reason for this, we can investigate how each circuit element behaves when either the voltage or current is a complex exponential. For the resistor, $v=Ri$. From our voltage given above, $i = \frac{V}{R}e^{j\omega t}$. Thus the resistor's voltage is a complex, as is the current with an amplitude $i = \frac{V}{R}$. For a capacitor, $i = C\frac{dv}{dt}$. Letting the voltage be a complex exponential we have $i = j\omega CVe^{j\omega t}$. The amplitude of this complex exponential is $I = j\omega CV$.

The major consequence of assuming complex exponential voltage and currents is that the ratio $Z = \frac{V}{I}$ for rather than depending on time each element depends on source frequency. This quantity is known as the element's (complex) impedance. The magnitude of the complex impedance is the ratio of the voltage amplitude to the current amplitude. Just like resistance in DC cases, impedance is the measure of the opposition that a circuit presents to the passage of a current when a voltage is applied. The impedance of a resistor is R, while that of a capacitor (C) is $\frac{1}{j\omega C}$. In the case of the circuit in, to find the complex impedance of the RC circuit, we add the impedance of the two components, just as with two resistors in series: $Z = R + \frac{1}{j\omega C}$.

Finding Real Currents and Voltages

Since $e^{j\omega t} = \cos(\omega t) + j\sin(\omega t)$, to find the real currents and voltages we simply need to take the real part of the *i(t)* and *v(t)*. The (real value) impedance is the real part of the complex impedance Z. For a series RC circuit, we get $Z = \sqrt{R^2 + \left(\frac{1}{\omega C}\right)^2}$. We see that the amplitude of the current will be $V/Z = \sqrt{R^2 + \left(\frac{1}{\omega C}\right)^2}$.

Phase Angle and Power Factor

In a series RC circuit connected to an AC voltage source, voltage and current maintain a phase difference.

Impedance is an AC (alternating current) analogue to resistance in a DC circuit. As we studied in a previously Atom ("Impedance"), current, voltage and impedance in an RC circuit are related by an AC version of Ohm 's law: $I = \frac{V}{Z}$, where I and V are peak current and peak voltage respectively, and Z is the impedance of the circuit.

In a series RC circuit connected to an AC voltage source as shown in figure, conservation of charge requires current be the same in each part of the circuit at all times. Therefore we can say: the currents in the resistor and capacitor are equal and in phase. (We will represent instantaneous current as *i(t)*.)

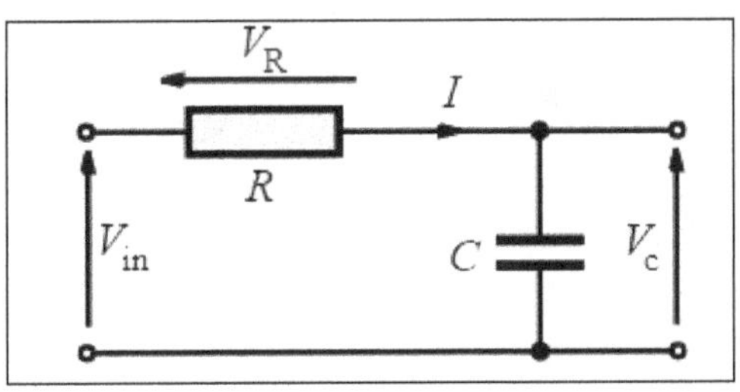

Series RC circuit.

On the other hand, because the total voltage should be equal to the sum of voltages on the resistor and capacitor, so we have:

$$v(t) = v_R(t) + v_C(t)$$
$$= i(t)R + i(t)\frac{1}{j\omega C},$$
$$= i(t)\left(R + \frac{1}{j\omega C}\right)$$

where, ω is the angular frequency of the AC voltage source and j is the imaginary unit; $j^2=-1$.

Since the complex number $Z = R + \frac{1}{j\omega C} = \sqrt{R^2 + \left(\frac{1}{\omega C}\right)^2}\, e^{j\phi}$ has a phase angle ϕ that satisfies $\cos\phi = \frac{R}{\sqrt{R^2 + \left(\frac{1}{\omega C}\right)^2}}$,

we notice that voltage v(t) and current i(t) has a phase difference of ϕ.

For R=0, ϕ=90°. As learned from the preceding series of Atoms—the voltage across the capacitor V_C follows the current by one-fourth of a cycle (or 90°).

Power Factor

Because voltage and current are out of phase, power dissipated by the circuit is not equal to: (*peak voltage*) times (*peak current*). The fact that source voltage and current are out of phase affects the power delivered to the circuit. It can be shown that the average power is $I_{rms}V_{rms}\cos\phi$, where I_{rms} and V_{rms} are the root mean square (rms) averages of the current and voltage, respectively. For this reason, $\cos\phi$ is called the power factor, which can range from 0 to 1.

Parallel Circuit

In a parallel circuit, there is more than one resistor (bulb) and they are arranged on many paths. This means electricity (electrons) can travel from one end of the cell through many branches to the other end of the cell.

Look at the figure below involving two resistors in a parallel circuit:

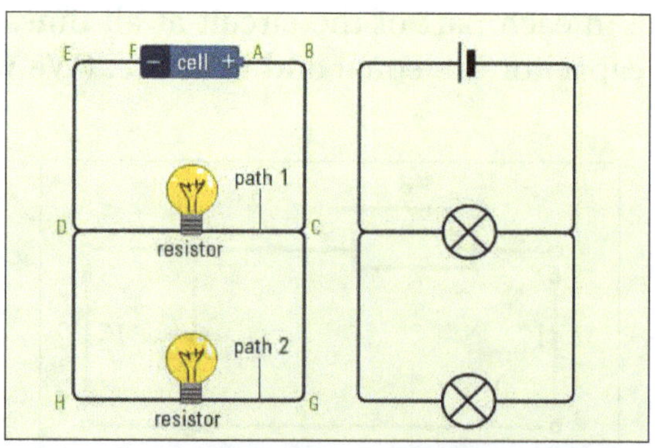

You will notice from the above that there is more than one path:

PATH 1: A-B-C-D-E-F and back to A

PATH 2: A-B-C-G-H-D-E-F and back to A

From the above, it is clear that electricity from the cell can take either path A or Path B to return to the cell. The great thing about parallel circuits is that, even when one resistor (bulb) burns out, the other bulbs will work because the electricity is not flowing through one path.

Think of all the light bulbs in your home. If one bulb burns out, the other bulbs in the rooms still work.

Another great thing is that the bulbs in a parallel circuit do not dim out like the case in series circuits. This is because the voltage across one branch is the same as the voltage across all other branches.

Circuit Elements

Circuit elements are those components of an electric circuit or electronic circuit which either can supply energy to the circuit or can take energy from the circuit. All the components of an circuit are connected together in required manner by conductors to form a desired and complete circuit. If the connecting conductor are taken as ideal, all the parts of the circuit, except the connecting conductors are circuit elements.

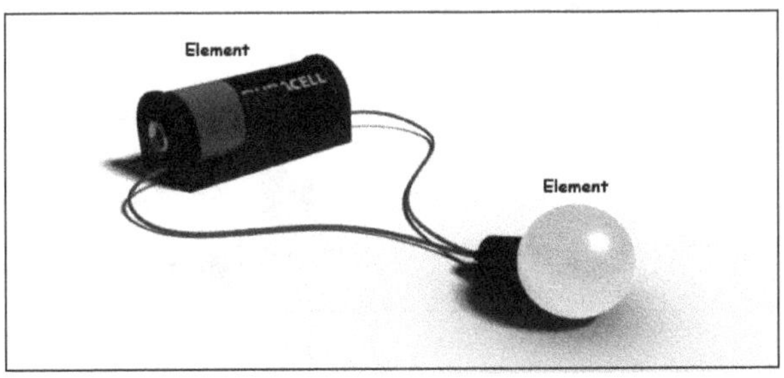

Types of Electric Circuit Elements

The electrical circuit elements either absorb energy from circuit or deliver energy to the circuit. Based on that there are two types of electric circuit elements, namely passive circuit elements and active circuit elements.

Passive Element

The circuit elements that receive energy (or absorb energy) and either convert it into heat or store it in an electric field or a magnetic field are called passive circuit elements.

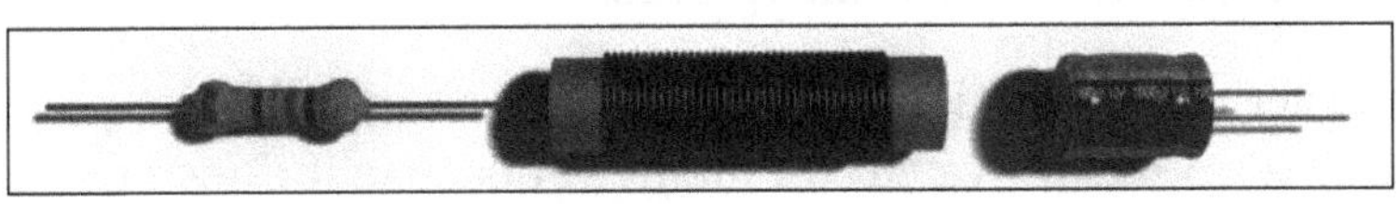

Resistor

Resistor is taken as passive element since it can dissipate energy as heat as long as current flows through it but in any situation a resistor cannot deliver energy to the circuit.

Inductor

An inductor is also considered as passive element of circuit, because it can store energy in it as magnetic field, and can deliver it to the circuit, but not in continuous basis. The energy absorbing and delivering capacity of an inductor are limited and transient in nature. That is why, an inductor is taken as passive element of a circuit.

Capacitor

For same reason, a capacitor is considered as passive element, because it can store energy in it as electric field and deliver it to the circuit, but not in continuous basis. The energy dealing capacity of a capacitor is limited and transient too.

Active Circuit Element

The circuit elements which supply energy to the circuit are called active circuit element.

Examples: Include voltage sources, current sources, and generators such as alternators, DC generators etc.

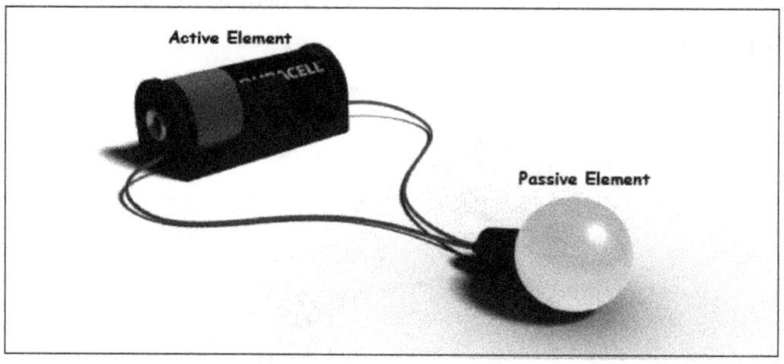

Voltage Source

A voltage source can be considered as an active element. If current leaves from the positive terminal of the voltage source, the energy is delivered from the source to the circuit.

As per definition of active element of electrical circuit, a battery can be considered as active element, as it continuously delivers energy to the circuit during discharging.

Current Source

A current source is also considered as active element. The current supplied to the circuit by an ideal current source is independent of circuit voltage.

Note: A transistor is an active circuit element, because, it can amplify the power of a signal. On the other hand,transformer is not an active circuit element because it does not amplify power and the power remains same both in primary and secondary side. Although, a transistor is an example of active circuit elements, but the transformer is an example of passive circuit elements.

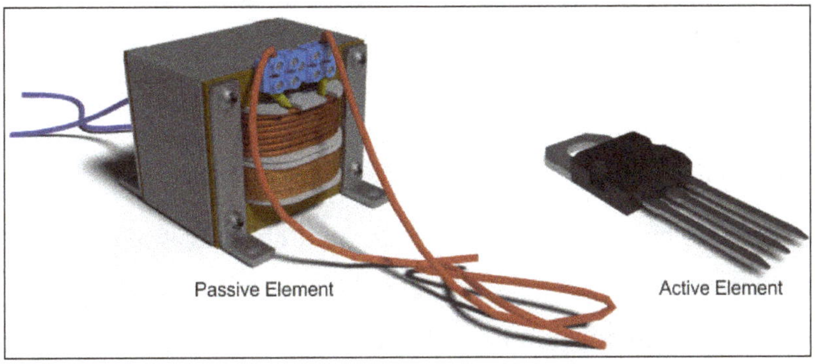

The circuit elements can also be categorised as bilateral and unilateral.

Bilateral Circuit Element

Conduction of current in both directions in a circuit element with same magnitude is termed as bilateral circuit element. It offers same resistance to the current of either directions.

Example: resistor, inductor, capacitor etc.

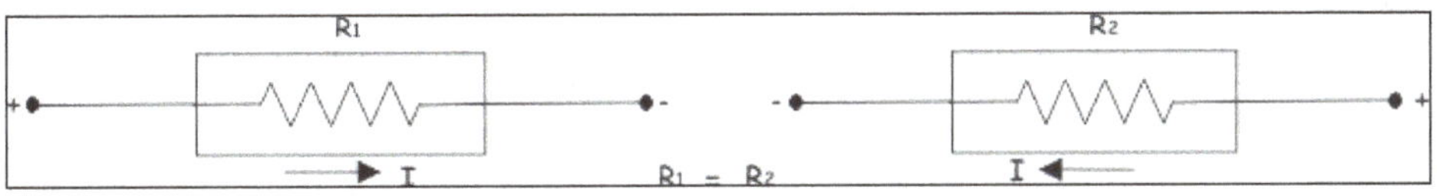

This figure shows a bilateral circuit element can conduct from both sides and offers same resistance for current from either side.

Unilateral Circuit Element

Conduction of current in one direction is termed as unilateral circuit element. It does not offer same resistance to the current of either direction. The resistance of the unilateral circuit element is different for forward current than that of reverse current.

Example: diode, transistor etc.

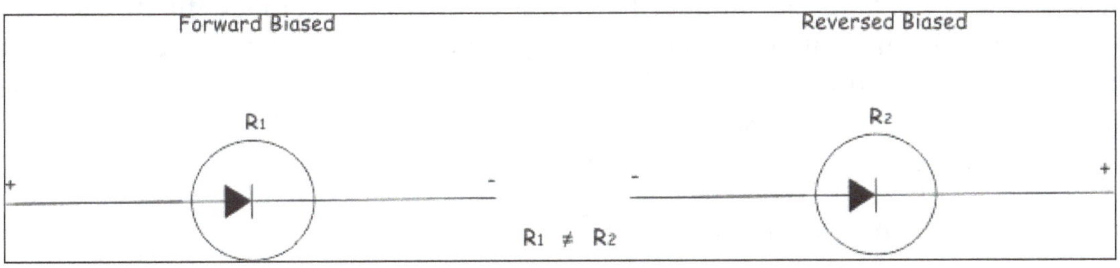

This figure shows, a diode as unilateral circuit element. When diode is forward biased it offers very small resistance and conducts. While it is reverse biased, it offers very high resistance and doesn't conduct. The circuit element can be categorized in another manner, such as lumped and distributed circuit elements.

Lumped Circuit Elements

When the voltage across and current through the element don't vary with dimension of the element, it is called lumped circuit elements.

Example: Resistor connected in any electrical circuit.

Distributed Circuit Element

When the voltage across and current through the element change with dimensions of the element, it is called distributed circuit element.

Example: Resistance of the transmission line. It varies with the length of the line.

Circuit Diagram

Circuit diagram is a simplified conventional graphical representation of an electrical circuit, showing how electrical components are connected together. It's usually used by engineers and electricians to explain symbolically parts and paths of an electrical circuit, which is important in design, construction, and maintenance of electrical and electronic equipment.

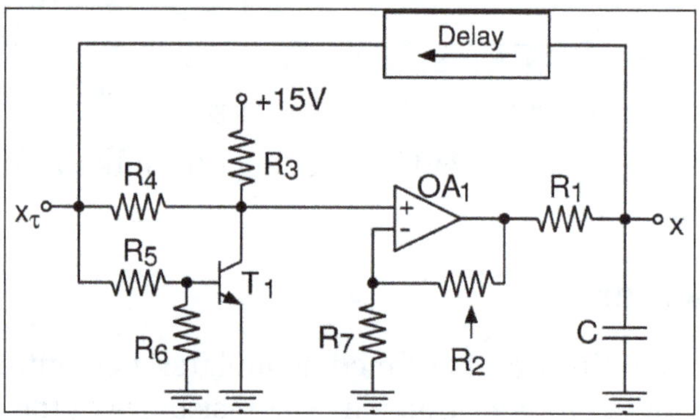

Standard Circuit Symbols for Designing

Circuit diagrams use international standard symbols to provide schematic diagrams of the circuit and its components. Each symbol is intended to represent some feature of the physical construction of the device. First you need to identify the symbols used to represent the components you wish to draw. Edraw provides you with lots of circuits symbols including cell, battery, source, resister, switch, lamp, motor, bell, buzzer, relay, capacitor, etc.

Below are some frequently used circuits symbols:

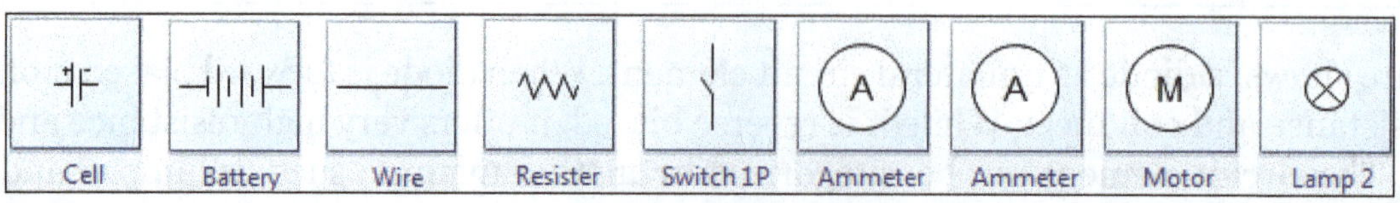

- Cell: A cell stands for electrical energy supply. A single cell is often called a battery, but strictly a battery is two or more cells joined together. It's represented by a long and a short parallel line.
- Battery: A battery is more than one cell. The larger terminal (on the left) is positive (+).It is represented by a collection of long and short parallel lines.
- Wire: Pass current from one part of the circuit to another. It's a connecting wire between two components.
- Resister: A resister is used to restrict the flow of a current. It's represented by a zigzag line.
- Switch: An on-off switch allows current to flow only when it is in the closed (on) position. It's generally represented by providing a break in a straight line by lifting a portion of the line upward at a diagonal.
- Ammeter: An ammeter is used to measure current. It's represented by a letter A in a circle.
- Voltmeter: A voltmeter is used to measure voltage. It's represented by a letter V in a circle.
- Motor: A transducer converts electrical energy to kinetic energy. It's represented by a letter M in a circle.
- Lamp: A transducer which converts electrical energy to light.

Types of Circuit Diagram

Circuit diagrams can be divided into two categories - pictorial circuit diagram and schematic circuit diagram.

Pictorial Circuit Diagram

Pictorial diagrams are much easier to understand than schematic circuit diagrams. Connecting realistic electrical components with the wiring, a pictorial diagram makes it easy for viewers to identify the electrical components of a system immediately without professional knowledge required.

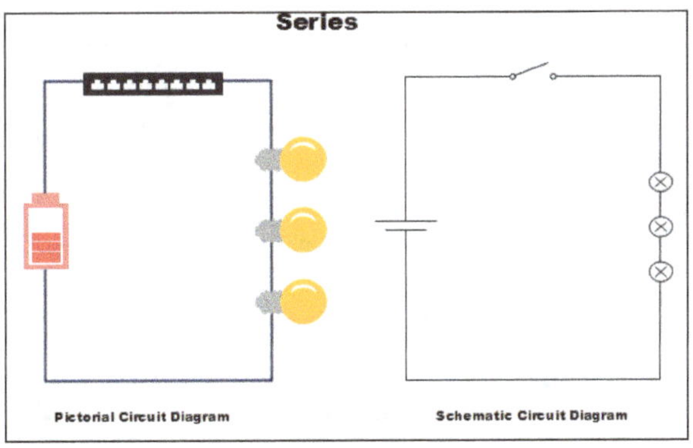

Schematic Circuit Diagram

A schematic circuit diagram represents the electrical system in the form of a diagram that shows the main features or relationships but not the details. In a schematic circuit diagram, the presentation of electrical components and wiring does not completely correspond to the physical arrangements in the real device. If you want to understand a schematic diagram, you are required to master basic knowledge of electricity and physics as well as internationally standardized symbols. Look at the parallel circuits below, you may find that the battery is represented as two short lines, the lights are represented as a circle with a cross inside, and the wiring is represented as a line. This kind of circuit diagram is mainly used by electrical engineers. The following is a schematic circuit diagram of a semiconductor electronic. Schematics and circuit diagrams are both important engineering diagrams.

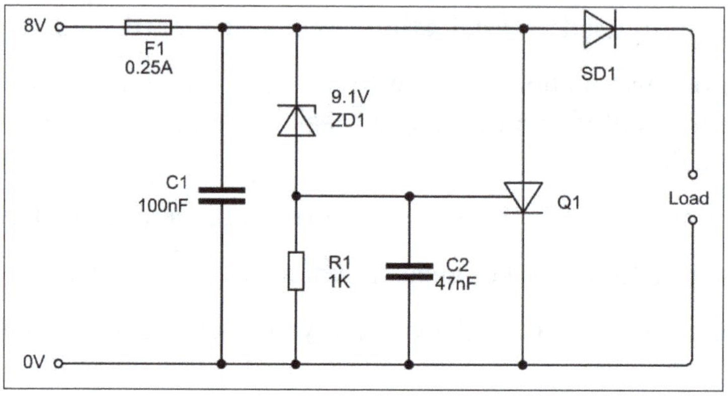

Phasor Diagrams

Phasor diagrams are a graphical way of representing the magnitude and directional relationship between two or more alternating quantities sinusoidal waveforms of the same frequency can have a Phase Difference between themselves which represents the angular difference of the two sinusoidal waveforms. Also the terms "lead" and "lag" as well as "in-phase" and "out-of-phase" are commonly used to indicate the relationship of one waveform to the other with the generalized sinusoidal expression given as: $A_{(t)} = A_m \, sin(\omega t \pm \Phi)$ representing the sinusoid in the time-domain form.

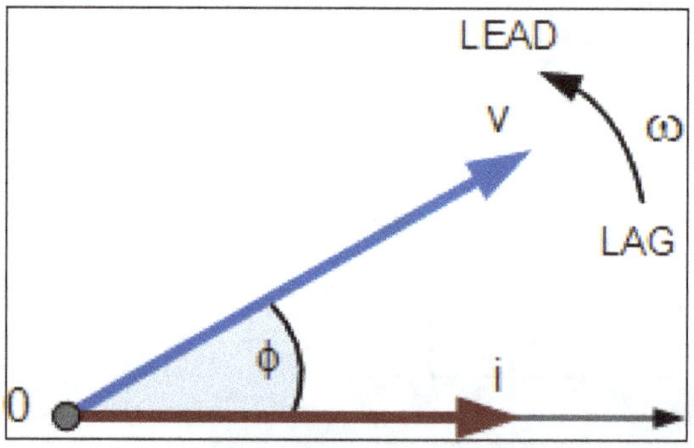

But when presented mathematically in this way it is sometimes difficult to visualise this angular or phasor difference between two or more sinusoidal waveforms. One way to overcome this problem is to represent the sinusoids graphically within the spacial or phasor-domain form by using Phasor diagrams, and this is achieved by the rotating vector method.

Basically a rotating vector, simply called a “Phasor” is a scaled line whose length represents an AC quantity that has both magnitude (“peak amplitude”) and direction (“phase”) which is “frozen” at some point in time.

A phasor is a vector that has an arrow head at one end which signifies partly the maximum value of the vector quantity (V or I) and partly the end of the vector that rotates.

Generally, vectors are assumed to pivot at one end around a fixed zero point known as the “point of origin” while the arrowed end representing the quantity, freely rotates in an anti-clockwise direction at an angular velocity, (ω) of one full revolution for every cycle. This anti-clockwise rotation of the vector is considered to be a positive rotation. Likewise, a clockwise rotation is considered to be a negative rotation.

Although the both the terms vectors and phasors are used to describe a rotating line that itself has both magnitude and direction, the main difference between the two is that a vectors magnitude is the “peak value” of the sinusoid while a phasors magnitude is the “rms value” of the sinusoid. In both cases the phase angle and direction remains the same.

The phase of an alternating quantity at any instant in time can be represented by a phasor diagram, so phasor diagrams can be thought of as “functions of time”. A complete sine wave can be constructed by a single vector rotating at an angular velocity of $\omega = 2\pi f$, where f is the frequency of the waveform. Then a Phasor is a quantity that has both “Magnitude” and “Direction”.

Generally, when constructing a phasor diagram, angular velocity of a sine wave is always assumed to be: ω in rad/sec.

Phasor Diagram of a Sinusoidal Waveform

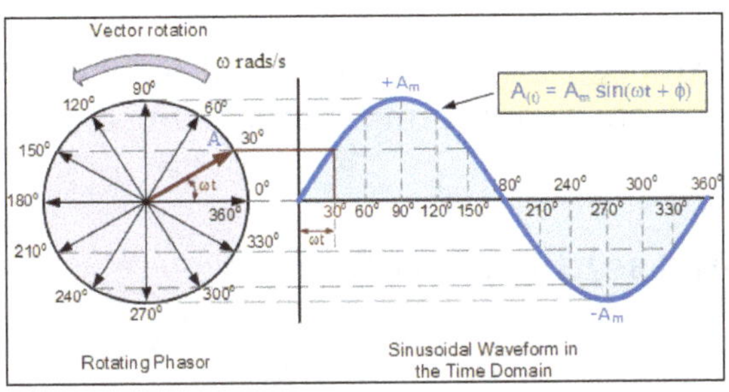

As the single vector rotates in an anti-clockwise direction, its tip at point A will rotate one complete revolution of 360° or 2π representing one complete cycle. If the length of its moving tip is transferred at different angular intervals in time to a graph as shown above, a sinusoidal waveform would be drawn starting at the left with zero time. Each position along the horizontal axis indicates the time that has elapsed since zero time, $t = 0$. When the vector is horizontal the tip of the vector represents the angles at 0°, 180° and at 360°.

Likewise, when the tip of the vector is vertical it represents the positive peak value, (+Am) at 90° or π/2 and the negative peak value, (-Am) at 270° or 3π/2. Then the time axis of the waveform represents the angle either in degrees or radians through which the phasor has moved. So we can say that a phasor represent a scaled voltage or current value of a rotating vector which is “frozen” at some point in time, (t) and in our example above, this is at an angle of 30°.

Sometimes when we are analysing alternating waveforms we may need to know the position of the phasor, representing the Alternating Quantity at some particular instant in time especially when we want to compare two different waveforms on the same axis. For example, voltage and current. We have assumed in the waveform above that the waveform starts at time t = 0 with a corresponding phase angle in either degrees or radians.

But if a second waveform starts to the left or to the right of this zero point or we want to represent in phasor notation the relationship between the two waveforms then we will need to take into account this phase difference, Φ of the waveform.

Phase Difference of a Sinusoidal Waveform

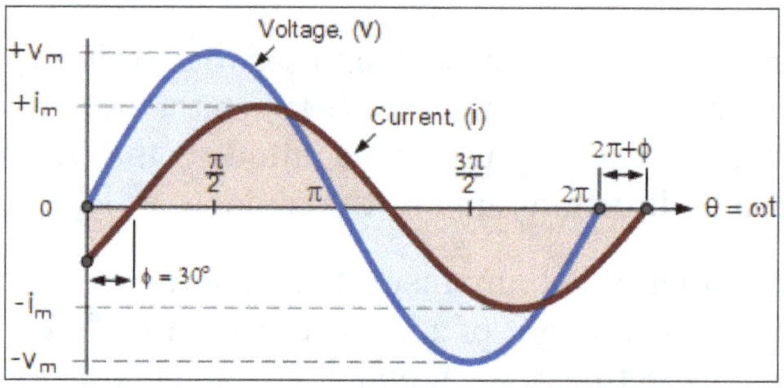

The generalised mathematical expression to define these two sinusoidal quantities will be written as:

$$v_{(t)} = V_m \sin(\omega t)$$
$$i_{(t)} = I_m \sin(\omega t - \phi)$$

The current, i is lagging the voltage, v by angle Φ and in our example above this is 30°. So the difference between the two phasors representing the two sinusoidal quantities is angle Φ and the resulting phasor diagram will be.

Phasor Diagram of a Sinusoidal Waveform

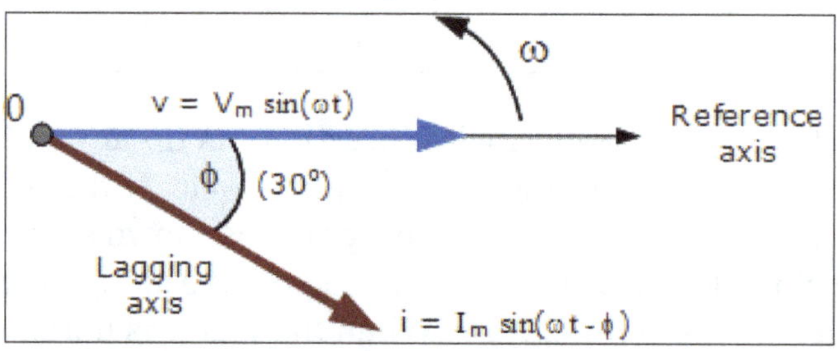

The phasor diagram is drawn corresponding to time zero (t=0) on the horizontal axis. The lengths of the phasors are proportional to the values of the voltage, (V) and the current, (I) at the instant in time that the phasor diagram is drawn. The current phasor lags the voltage phasor by the angle, Φ, as the two phasors rotate in an *anticlockwise*direction as stated earlier, therefore the angle, Φ is also measured in the same anticlockwise direction.

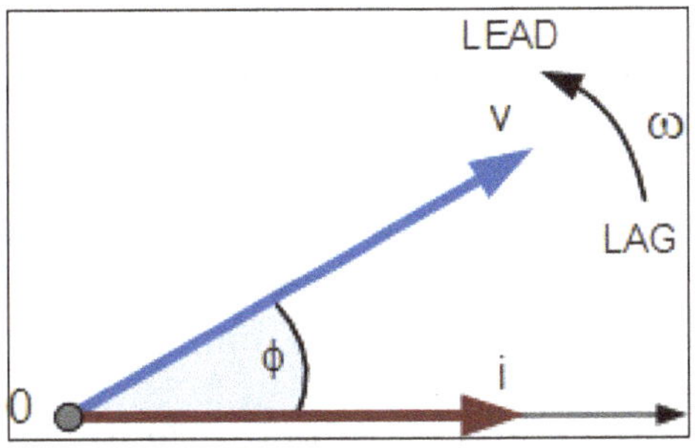

If however, the waveforms are frozen at time, t = 30°, the corresponding phasor diagram would look like the one shown on the right. Once again the current phasor lags behind the voltage phasor as the two waveforms are of the same frequency.

However, as the current waveform is now crossing the horizontal zero axis line at this instant in time we can use the current phasor as our new reference and correctly say that the voltage phasor is "leading" the current phasor by angle, Φ. Either way, one phasor is designated as the *reference* phasor and all the other phasors will be either leading or lagging with respect to this reference.

Phasor Addition

Sometimes it is necessary when studying sinusoids to add together two alternating waveforms, for example in an AC series circuit, that are not in-phase with each other. If they are in-phase that is, there is no phase shift then they can be added together in the same way as DC values to find the algebraic sum of the two vectors. For example, if two voltages of say 50 volts and 25 volts respectively are together "in-phase", they will add or sum together to form one voltage of 75 volts (50 + 25).

If however, they are not in-phase that is, they do not have identical directions or starting point then the phase angle between them needs to be taken into account so they are added together using phasor diagrams to determine their Resultant Phasor or Vector Sum by using the *parallelogram law*.

Consider two AC voltages, V_1 having a peak voltage of 20 volts, and V_2 having a peak voltage of 30 volts where V_1 leads V_2 by 60°. The total voltage, V_T of the two voltages can be found by firstly drawing a phasor diagram representing the two vectors and then constructing a parallelogram in which two of the sides are the voltages, V_1 and V_2 as shown below.

Phasor Addition of Two Phasors

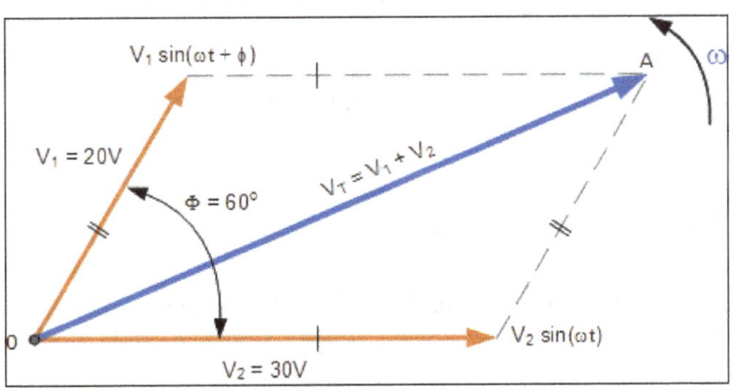

By drawing out the two phasors to scale onto graph paper, their phasor sum $V_1 + V_2$ can be easily found by measuring the length of the diagonal line, known as the "resultant r-vector", from the zero point to the intersection of the construction lines 0-A. The downside of this graphical method is that it is time consuming when drawing the phasors to scale.

Also, while this graphical method gives an answer which is accurate enough for most purposes, it may produce an error if not drawn accurately or correctly to scale. Then one way to ensure that the correct answer is always obtained is by an analytical method.

Mathematically we can add the two voltages together by firstly finding their "vertical" and "horizontal" directions, and from this we can then calculate both the "vertical" and "horizontal" components for the resultant "r vector", V_T. This analytical method which uses the cosine and sine rule to find this resultant value is commonly called the Rectangular Form.

In the rectangular form, the phasor is divided up into a real part, x and an imaginary part, y forming the generalised expression Z = x ± jy. This then gives us a mathematical expression that represents both the magnitude and the phase of the sinusoidal voltage as:

Complex Sinusoid

$$V_m = \cos(\phi) + jV_m(\sin\phi)$$

So the addition of two vectors, A and B using the previous generalised expression is as follows:

$$A = x + jy \quad and \quad B = w + jz$$

$$A + B = (x + w) + j(y + z)$$

Phasor Addition using Rectangular Form

Voltage, V_2 of 30 volts points in the reference direction along the horizontal zero axis, then it has a horizontal component but no vertical component as follows:

- Horizontal Component = 30 cos 0° = 30 volts.
- Vertical Component = 30 sin 0° = 0 volts.

This then gives us the rectangular expression for voltage V_2 of: 30 + j0

Voltage, V_1 of 20 volts leads voltage, V_2 by 60°, then it has both horizontal and vertical components as follows:

- Horizontal Component = 20 cos 60° = 20 x 0.5 = 10 volts.
- Vertical Component = 20 sin 60° = 20 x 0.866 = 17.32 volts.

This then gives us the rectangular expression for voltage V_1 = 10 + j17.32

The resultant voltage, V_T is found by adding together the horizontal and vertical components as follows:

$V_{Horizontal}$ = sum of real parts of V_1 and V_2 = 30 + 10 = 40 volts

$V_{Vertical}$ = sum of imaginary parts of V_1 and V_2 = 0 + 17.32 = 17.32 volts.

Now that both the real and imaginary values have been found the magnitude of voltage, V_T is determined by simply using Pythagoras's Theorem for a 90° triangle as follows:

$$V_T = \sqrt{\left(\begin{matrix}\text{Real or Horizontal}\\ \text{Component}\end{matrix}\right)^2 + \left(\begin{matrix}\text{Imaginary or Vertical}\\ \text{Component}\end{matrix}\right)^2}$$

$$V_T = \sqrt{40^2 + 17.32^2}$$

$$\therefore V_T = 43.6 volts$$

Then the resulting phasor diagram will be,

Resultant Value of V_T

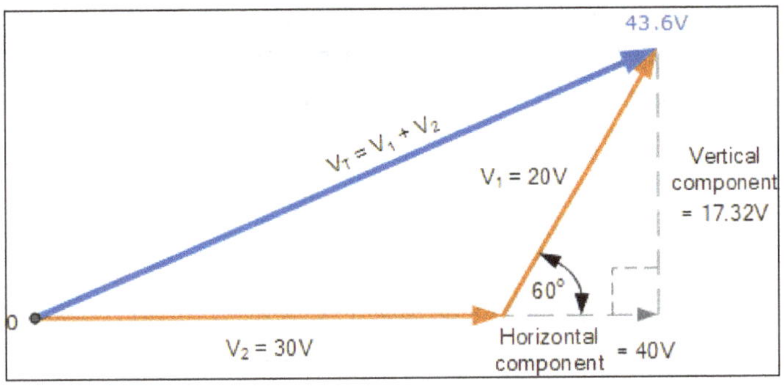

Phasor Subtraction

Phasor subtraction is very similar to the above rectangular method of addition, except this time the vector difference is the other diagonal of the parallelogram between the two voltages of V_1 and V_2 as shown.

Vector Subtraction of Two Phasors

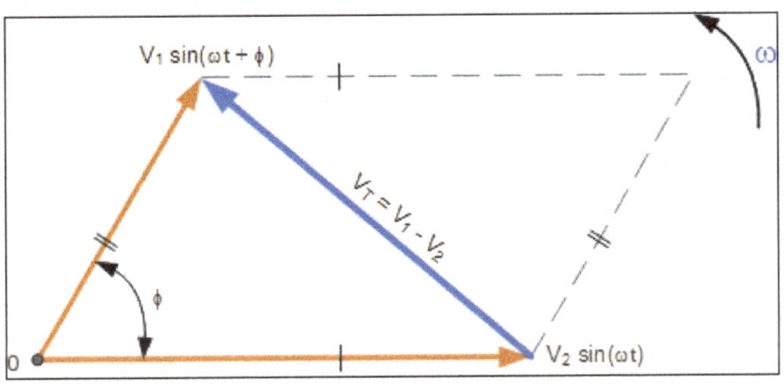

This time instead of "adding" together both the horizontal and vertical components we take them away, subtraction.

$$A = x + jy \quad and \quad B = w + jz$$

$$A - B = (x - w) + j(y - z)$$

The 3-phase Phasor Diagram

Previously we have only looked at single-phase AC waveforms where a single multi-turn coil rotates within a magnetic field. But if three identical coils each with the same number of coil turns are placed at an electrical angle of 120° to each other on the same rotor shaft, a three-phase voltage supply would be generated.

A balanced three-phase voltage supply consists of three individual sinusoidal voltages that are all equal in magnitude and frequency but are out-of-phase with each other by exactly 120° electrical degrees.

Standard practice is to colour code the three phases as Red, Yellow and Blue to identify each individual phase with the red phase as the reference phase. The normal sequence of rotation for a three phase supply is Red followed by Yellow followed by Blue, (R, Y, B).

As with the single-phase phasors above, the phasors representing a three-phase system also rotate in an anti-clockwise direction around a central point as indicated by the arrow marked ω in rad/s. The phasors for a three-phase balanced star or delta connected system are shown below.

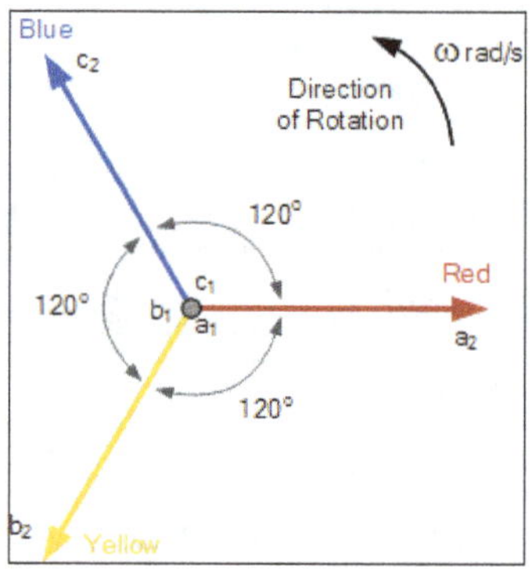

The phase voltages are all equal in magnitude but only differ in their phase angle. The three windings of the coils are connected together at points, a_1, b_1 and c_1 to produce a common neutral connection for the three individual phases. Then if the red phase is taken as the reference phase each individual phase voltage can be defined with respect to the common neutral.

Three-phase Voltage Equations

$$\text{Red}\,Phase\text{: } V_{RN} = V_m \sin\theta$$

$$Yellow\,Phase\text{: } V_{YN} = V_m \sin\left(\theta - 120^0\right)$$

$$Blue\,Phase \quad V_{BN} = V_m \sin\left(\theta - 240^0\right)$$

$$or$$

$$V_{RN} = V_m \sin\left(\theta + 120^0\right)$$

If the red phase voltage, V_{RN} is taken as the reference voltage as stated earlier then the phase sequence will be R – Y – B so the voltage in the yellow phase lags V_{RN} by 120°, and the voltage in the blue phase lags V_{YN} also by 120°. But we can also say the blue phase voltage, V_{BN} leads the red phase voltage, V_{RN} by 120°.

One final point about a three-phase system. As the three individual sinusoidal voltages have a fixed relationship between each other of 120° they are said to be "balanced" therefore, in a set of balanced three phase voltages their phasor sum will always be zero as: $V_a + V_b + V_c = 0$.

Electrical Impedance

Electrical impedance (also known as "impedance" in short) is an addition of the definition of resistance to analternating current(AC). Which means that impedance includes both resistance (opposition of the electric current that causes heat) and reactance (a measure of such an oppositional current alternates) – in detail, the opposition adjacent to the electric currents. In the direct current (DC), electrical impedance is same as a resistance, except that it does not hold true in AC circuits.

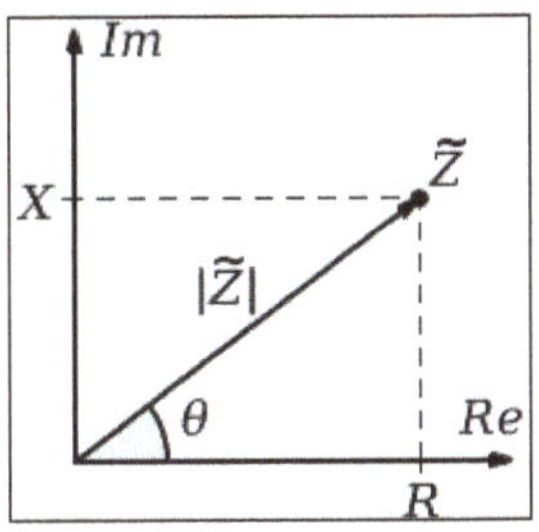

Electrical Impedance.

Impedance can also be dissimilar from resistance when a DC circuit changes flow in one way or other- similar to the opening and closing of an electrical switch, as is observed in the computers when they open and close switches to represent ones and zeros (binary language). The opposite of an impedance is admittance, which is the measure of the allowance of current. The figure to the left is a complex impedance plane, in which impedance is represented by a Z, resistance is depicted as R, and the reactance is depicted with X.

Electrical Impedance Tomography (EIT)

The fundamental principle of electrical impedance tomography (EIT) is akin to electrical resistance tomography (ERT) such that several measurements at the periphery of a process vessel or tube are taken and combined to give information on the electrical properties of the process volume.

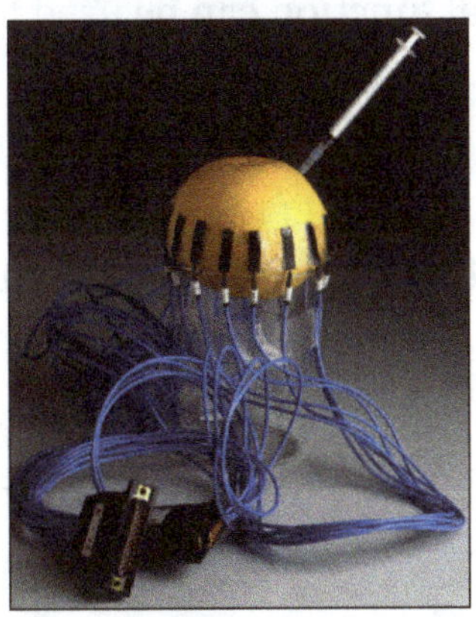

Electrical Impedance Tomography.

Electrical Impedance Tomography (EIT) is a non-invasive medical imaging method in which a figure of the conductivity or permitivity of a part of the body is incidental from the surface electrode measurements. Electrical conductivity depends on the free ion content and differs significantly between different biological tissues (absolute EIT) or dissimilar practical states of one and other similar tissues or organs (relative or functional EIT). Majority of EIT systems apply little irregular currents at a single frequency; however, some EIT systems use various frequencies to better discriminate between usual and suspected abnormal tissue within the same organ (multifrequency-EIT or electrical impedance spectroscopy).

Complex Impedance

A resistor with a value of R has an impedance of R ohms, a real number. An ideal inductor has a complex impedance of,

$$Z = j2\pi L$$

Where, 'f' is the frequency in Hertz and L is the inductance in Henries. It is imaginary because an a ideal inductor can simply store and release electrical energy. It can't dissipate it as heat like a resistor. Similarly, an ideal capacitor has a complex impedance of,

$$Z = \frac{-j}{2\pi fc}$$

Where 'C' is the capacitance in farads.

Use of Complex Impedance

The behavior of the impedance of an AC circuit with various components quickly becomes unmanageable if sines and cosines are used to present the voltages and current. A mathematical build which eases the complexity use of complex exponential functions. The necessary parts of the strategy are as follows.

Math relationship underlaying the technique,

$$ej\omega t = cos\omega t + sin\omega t$$

The real part of a complex exponential function can be used to represent an AC voltage or current.

$$V = Vm\ COS\omega t$$
$$I = Im\ COS\ (\omega t - \varphi)$$

The impedance can then be expressed as a complex exponential,

$$Z = Vm /\ Im\ e - j\emptyset = R + jX$$

The impedance of the individual circuit elements can then be expressed as pure real or imaginary numbers.

$$R\ - j / \omega c\ j\omega L$$

Complex Impedance for RL and RC

Using complex impedance is a significant technique for handling multi-component AC circuits. If a complex plane is used with resistance along the real axis, then the reactance of a capacitor and inductor are treated as imaginary numbers. For series combinations of the components such as RL and RC combinations, the component values are added as if they were components of a vector. Shown now is the Cartesian form of the complex impedance. They can also be written in polar form. Impedances in combination circuits like the RLC parallel circuit.

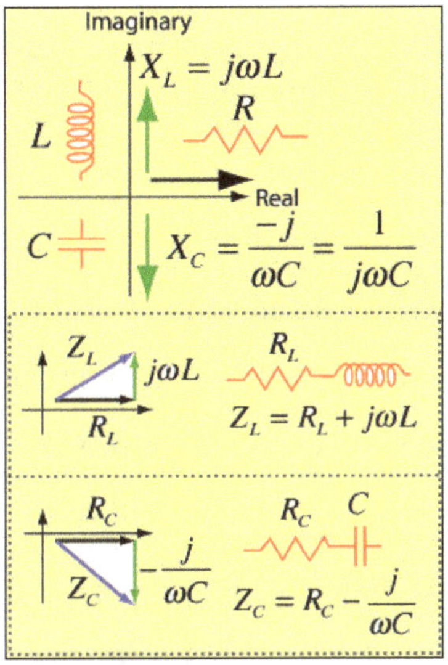

Complex Impedance for RL and RC.

Resistance and Reactance

Resistance is fundamentally a friction against the motion of electrons. It is there in all conductors to some extent (except superconductors!), and most notably in resistors. When alternating current goes through a resistance, a voltage drop is formed that is in-phase with the current. Resistance is mathematically symbolized by the letter "R" and is measured in the unit of ohms (Ω).

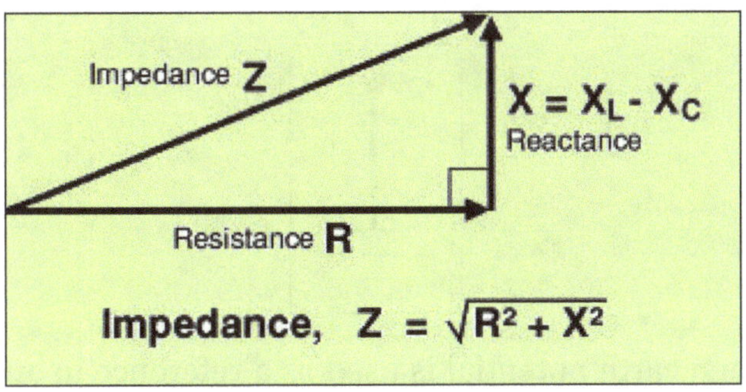

Resistance and Reactance Circuit.

Reactance is essentially inactive against the motion of electrons. It is present anywhere electric or magnetic fields are developed in proportion to an applied voltage or current, correspondingly; but most notably in capacitors and inductors. When alternating current goes through a pure reactance, a voltage drop is produced, which is 900 out of phase with the current. Reactance is a mathematically symbolized by the letter "X" and is measured in the unit of Ohms (Ω).

Network Analysis

Network Analysis for electric circuits are the different useful techniques related to several currents, emfs, and resistance voltages in such circuit. This is somewhat the collection of techniques of finding the voltages and currents in every component of the network.

Nodal Analysis

Nodal analysis is a method that provides a general procedure for analyzing circuits using node voltages as the circuit variables. Nodal Analysis is also called the Node-Voltage Method.

Some features of nodal analysis are as follows:

- Nodal Analysis is based on the application of the Kirchhoff's Current Law (KCL).
- Having 'n' nodes there will be 'n-1' simultaneous equations to solve.
- Solving 'n-1' equations all the nodes voltages can be obtained.
- The number of non reference nodes is equal to the number of Nodal equations that can be obtained.

Types of Nodes in Nodal Analysis

- Non Reference Node: It is a node which has a definite Node Voltage. e.g. Here Node 1 and Node 2 are the Non Reference nodes.
- Reference Node: It is a node which acts a reference point to all the other node. It is also called the Datum Node.

Types of Reference Nodes

- Chassis Ground: This type of reference node acts a common node for more than one circuits.

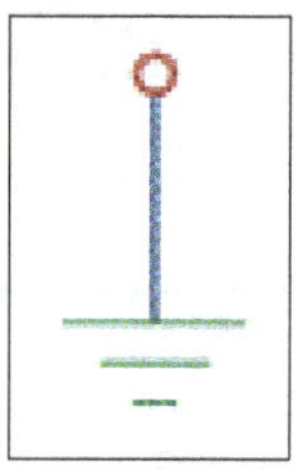

- Earth Ground: When earth potential is used as a reference in any circuit then this type of reference node is called Earth Ground.

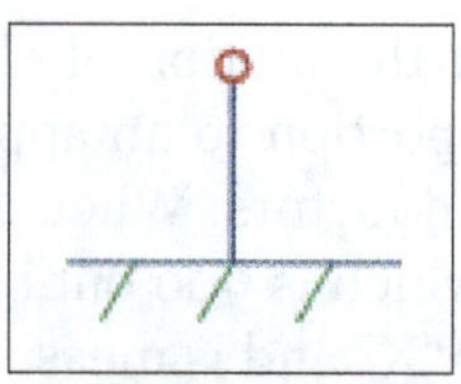

Solving of Circuit using Nodal Analysis

Basic Steps used in Nodal Analysis

- Select a node as the reference node. Assign voltages V_1, V_2... V_{n-1} to the remaining nodes. The voltages are referenced with respect to the reference node.
- Apply KCL to each of the non reference nodes.
- Use Ohm's law to express the branch currents in terms of node voltages.

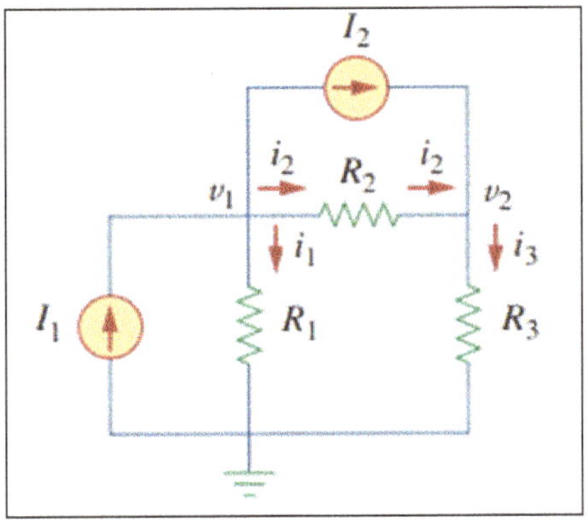

Node always assumes that current flows from a higher potential to a lower potential in resistor. Hence, current is expressed as follows,

$$I = \frac{V_{high} - V_{low}}{R}$$

- After the application of Ohm's Law get the 'n-1' node equations in terms of node voltages and resistances.
- Solve 'n-1' node equations for the values of node voltages and get the required node Voltages as result.

Nodal Analysis with Current Sources

Nodal analysis with current sources is very easy and it is discussed with a example below.

Example: Calculate Node Voltages in following circuit,

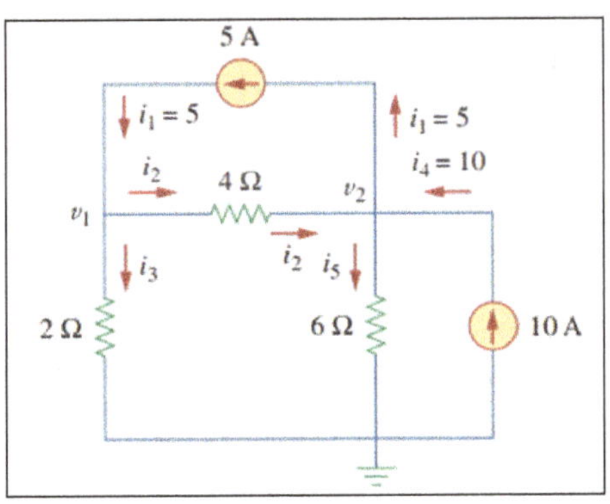

In the following circuit we have 3 nodes from which one is reference node and other two are non reference nodes - Node 1 and Node 2.

Step I. Assign the nodes voltages as v_1 and v_2 and also mark the directions of branch currents with respect to the reference nodes.

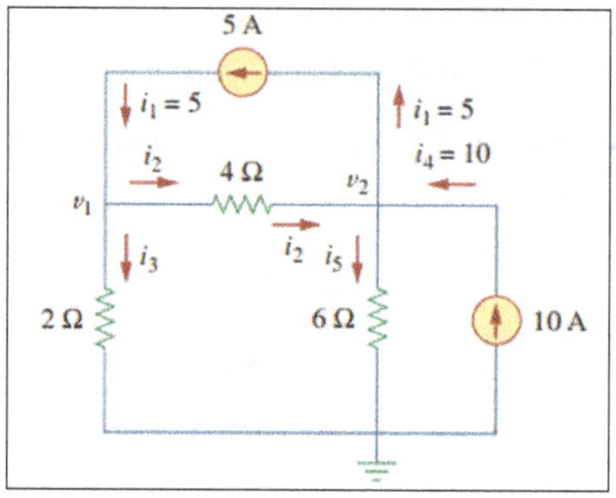

Step II. Apply KCL to Nodes 1 and 2.

KCL at Node 1,

$$i_1 = i_2 + i_3$$

KCL at Node 2,

$$i_2 + i_4 = i_1 + i_5$$

Step III. Apply Ohm's Law to KCL equations,

- Ohm's law to KCL equation at Node 1,

$$i_1 = i_2 + i_3 \Rightarrow 5 = \frac{v_1 - v_2}{4} + \frac{v_2 - 0}{4}$$

Simplifying the above equation we get,

$$3v_1 - v_2 = 20$$

- Now, Ohm's Law to KCL equation at Node 2,

$$i_2 + i_4 = i_1 + i_5 \Rightarrow \frac{v_1 - v_2}{4} + 10 = 5 + \frac{v_2 - 0}{6}$$

Simplifying the above equation we get,

$$-3v_1 + 5v_2 = 60$$

Step IV. Now solve the above two equations to get the values of v_1 and v_2 as,

Using elimination method,

$$3v_1 - v_2 = 20$$
$$-3v_1 + 5v_2 = 60$$
$$\Rightarrow 4v_2 = 80 \Rightarrow v_2 = 20\,Volts$$

And substituting value v_2 = 20 Volts in equation mentioned before the above one, we get,

$$3v_1 - 20 = 20 \Rightarrow v_1 = \frac{40}{3} = 13.333\,Volts$$

Hence node voltages are as v1 = 13.33 Volts and v2 = 20 Volts.

Nodal Analysis with Voltage Sources

Case I. If a voltage source is connected between the reference node and a non reference node, we simply set the voltage at the non-reference node equal to the voltage of the voltage source and its analysis can be done as we done with current sources. v_1 = 10 Volts.

Case II. If the voltage source is between the two non reference nodes then it forms a supernode whose analysis is done as following:

Supernode Analysis

Whenever a voltage source (Independent or Dependent) is connected between the two non reference nodes then these two nodes form a generalized node called the Super node. So, Super node can be regarded as a surface enclosing the voltage source and its two nodes.

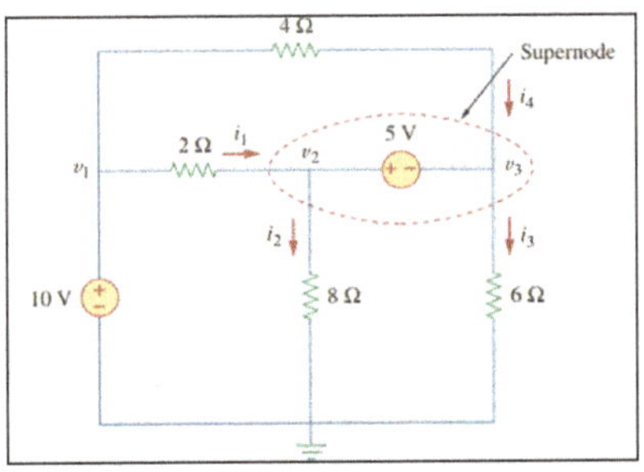

In the above figure source is connected between two non reference nodes Node - 2 and Node - 3. So here Node - 2 and Node - 3 form the Super node.

Properties of Supernode

- Always the difference between the voltage of two non reference nodes is known at Supernode.
- A supernode has no voltage of its own.
- A supernode requires application of both KCL and KVL to solve it.
- Any element can be connected in parallel with the voltage source forming the supernode.
- A Supernode satisfies the KCL as like a simple node.

Solving Any Circuit Containing Supernode

Let's take a example to understand how to solve circuit containing Supernode.

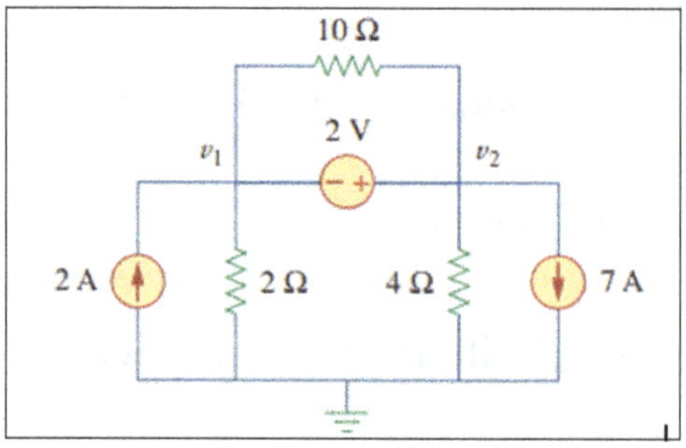

Here 2V voltage source is connected between Node-1 and Node-2 and it forms a Supernode with a 10Ω resistor in parallel.

Note - Any element connected in parallel with the voltage source forming Super node doesn't make any difference because $v_2 - v_1 = 2V$ always whatever may be the value of resistor. Thus 10 Ω can be removed and circuit is redrawn and applying KCL to the supernode as shown in figure gives,

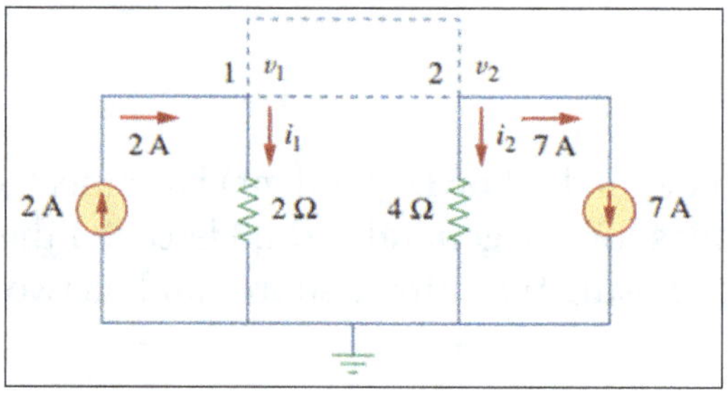

Expressing and in terms of the node voltages.

$$2 = \frac{v_1 - 0}{2} + \frac{v_2 - 0}{4} + 7 \Rightarrow 8 = 2v_1 + v_2 + 28 \Rightarrow v_2 = -2v_1 - 20$$

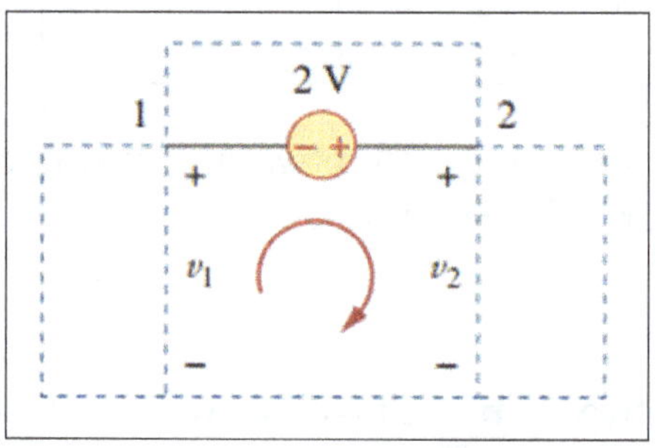

$$v_1 + 2 - v_2 = 0 \Rightarrow v_2 = v_1 + 2$$

From above two equations we can write as,

$$v_2 = v_1 + 2 = -2v_1 - 20 \Rightarrow 3v_1 = -22$$
$$\Rightarrow v_1 = -7.333V \;\&\; v_2 = v_1 + 2 = -7.333 + 2 = -5.333V$$

Hence, v_1 = - 7.333V and v_2 = - 5.333V which is required answer.

Mesh Analysis

Mesh Current Analysis is a technique used to find the currents circulating around a loop or mesh with in any closed path of a circuit.

Mesh Current Analysis Circuit

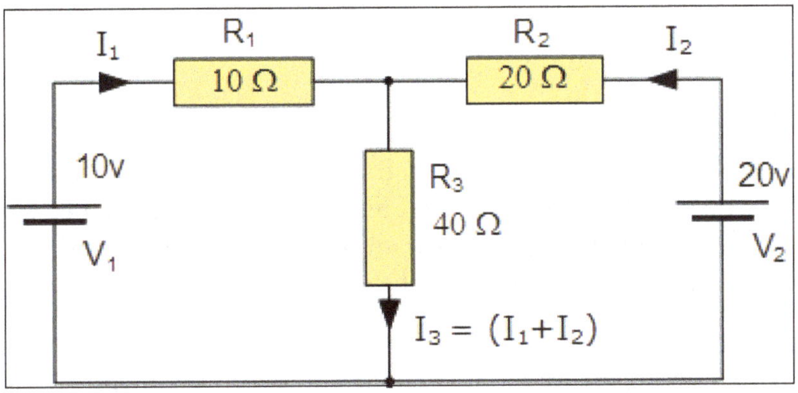

One simple method of reducing the amount of math's involved is to analyse the circuit using Kirchhoff's Current Law equations to determine the currents, I_1 and I_2 flowing in the two resistors. Then there is no need to calculate the current I_3 as its just the sum of I_1 and I_2. So Kirchhoff's second voltage law simply becomes:

$$10 = 50I_1 + 40I_2$$

$$20 = 40I_1 + 60I_2$$

therefore, one line of math's calculation have been saved.

Mesh Current Analysis

An easier method of solving the above circuit is by using Mesh Current Analysis or Loop Analysis which is also sometimes called Maxwell´s Circulating Currents method. Instead of labelling the branch currents we need to label each "closed loop" with a circulating current.

As a general rule of thumb, only label inside loops in a clockwise direction with circulating currents as the aim is to cover all the elements of the circuit at least once. Any required branch current may be found from the appropriate loop or mesh currents as before using Kirchhoff´s method.

For example: $i_1 = I_1$, $i_2 = -I_2$ and $I_3 = I_1 - I_2$

We now write Kirchhoff's voltage law equation in the same way as before to solve them but the advantage of this method is that it ensures that the information obtained from the circuit equations is the minimum required to solve the circuit as the information is more general and can easily be put into a matrix form.

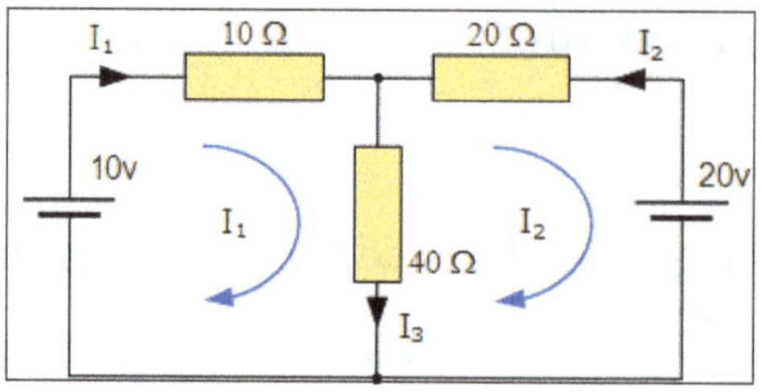

$10=50I_1-40I_2$

$-20=.40I_1+60I_2$

or [V]=[R][I]

These equations can be solved quite quickly by using a single mesh impedance matrix Z. Each element ON the principal diagonal will be "positive" and is the total impedance of each mesh. Where as, each element OFF the principal diagonal will either be "zero" or "negative" and represents the circuit element connecting all the appropriate meshes.

First we need to understand that when dealing with matrices, for the division of two matrices it is the same as multiplying one matrix by the inverse of the other as shown.

$$[V]=[I]\times[R] \ or \ [R]\times[I]=[V]$$

$$\begin{bmatrix} 50 & -40 \\ -40 & 60 \end{bmatrix} \times \begin{bmatrix} I_1 \\ I_2 \end{bmatrix} = \begin{bmatrix} 10 \\ -20 \end{bmatrix}$$

$$I = \frac{V}{R} = R^{-1} \times V$$

$$Inverse \ of \ R = \begin{bmatrix} 60 & 40 \\ 40 & 50 \end{bmatrix}$$

$$|R| = (60 \times 50) - (40 \times 40) = 1400$$

$$\therefore R^{-1} = \frac{1}{1400} \begin{bmatrix} 60 & 40 \\ 40 & 50 \end{bmatrix}$$

Having found the inverse of R, as V/R is the same as V x R^{-1}, we can now use it to find the two circulating currents.

$$[I] = [R^{-1}] \times [V]$$

$$\begin{bmatrix} I_1 \\ I_2 \end{bmatrix} = \frac{1}{1400} \begin{bmatrix} 60 & 40 \\ 40 & 50 \end{bmatrix} \times \begin{bmatrix} 10 \\ -20 \end{bmatrix}$$

$$I_1 = \frac{(60\times10)+(40\times-20)}{1400} = \frac{-200}{1400} = -0.143A$$

$$I_2 = \frac{(40\times10)+(50\times-20)}{1400} = \frac{-600}{1400} = -0.429A$$

Where,

[V] gives the total battery voltage for loop 1 and then loop 2.

[I] states the names of the loop currents which we are trying to find.

[R] is the resistance matrix.

[R^{-1}] is the inverse of the [R] matrix.

and this gives I_1 as -0.143 Amps and I_2 as -0.429 Amps

$I_3 = I_1 - I_2$

The combined current of I_3 is therefore given as : -0.143 – (-0.429) = 0.286 Amps

which is the same value of 0.286 amps.

Ohm's Law

Ohm's Law states that the current flowing in a circuit is directly proportional to the applied potential difference and inversely proportional to the resistance in the circuit.

In other words by doubling the voltage across a circuit the current will also double. However if the resistance is doubled the current will fall by half.

In this mathematical relationship the unit of resistance is measured in Ohms.

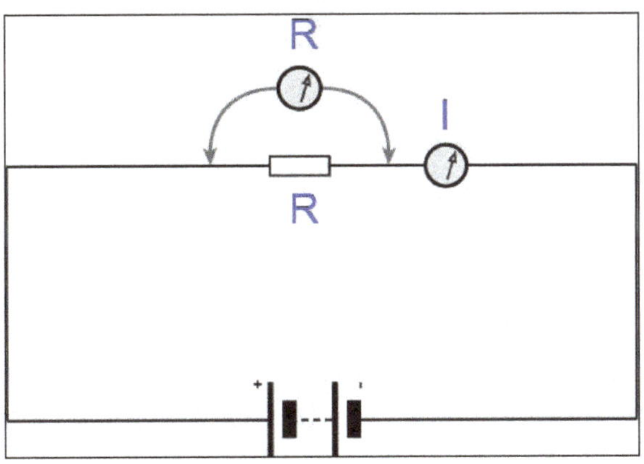

Ohm's Law Formula

The Ohm's Law formula or equation is very straightforward.

Ohm's law can be expressed in a mathematical form:

$$V = IR$$

Where,

V = voltage expressed in Volts

I = current expressed in Amps

R = resistance expressed in Ohms

The formula can be manipulated so that if any two quantities are known the third can be calculated.

$$I = \frac{V}{R}$$

$$R = \frac{V}{I}$$

Ohm's Law Triangle

To help remember the formula it is possible to use a triangle with one side horizontal and the peak at the top like a pyramid. This is sometimes known as the Ohm's law triangle.

In the top corner of the Ohm's law triangle is the letter V, in the left hand corner, the letter I, and in the right hand bottom corner, R.

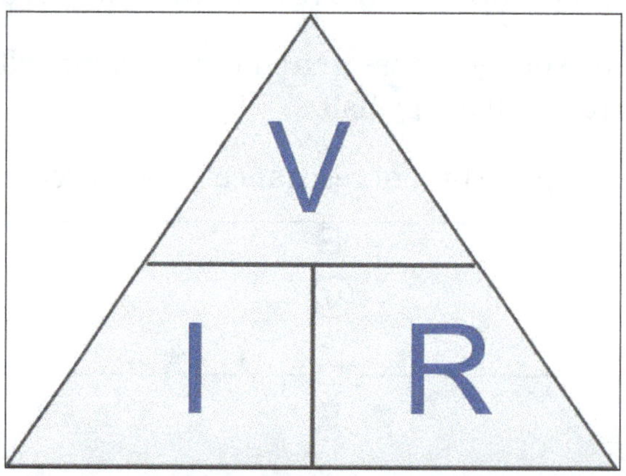

To use the triangle cover up the unknown quantity and then and then calculate it from the other two. If they are in line they are multiplied, but if one is on top of the other then they should be divided. In other words if current has to be calculated the voltage is divided by the resistance i.e. V/R and so forth.

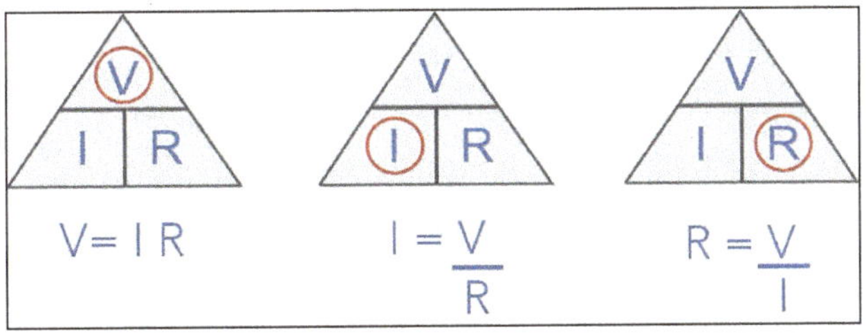

If the voltage has to be calculated then it is found by multiplying the current by the resistance i.e. I x R.

Ohm's Law Calculation Example

If a voltage of 10 volts is placed across a 500 ohm resistor determine the amount of current that will flow.

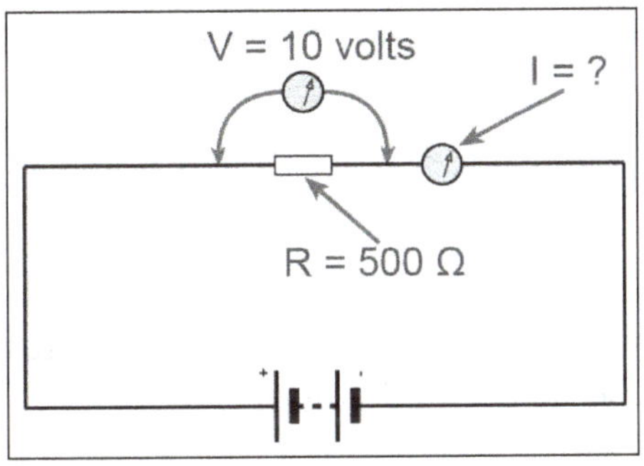

Looking at the Ohm's Law triangle the current is the unknown leaving the voltage and resistance as the known values.

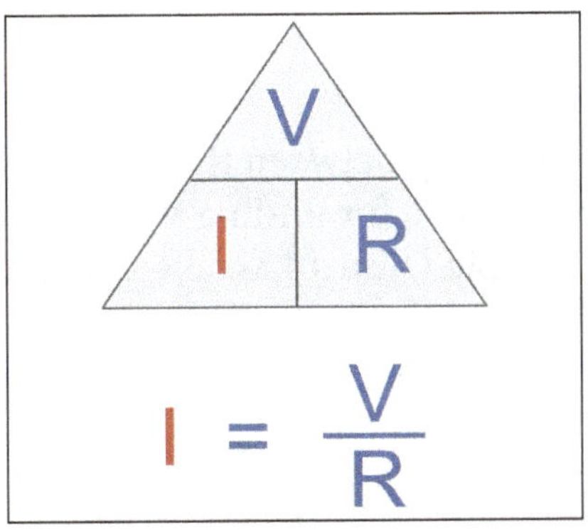

In this way the current is found by dividing the voltage by the resistance.

$$I = \frac{V}{R} = \frac{10}{500} = 0.02A = 20mA$$

Linear Graph

It can be seen that if the voltage and current were plotted for a fixed resistor or a length of wire, etc, there would be a linear curve.

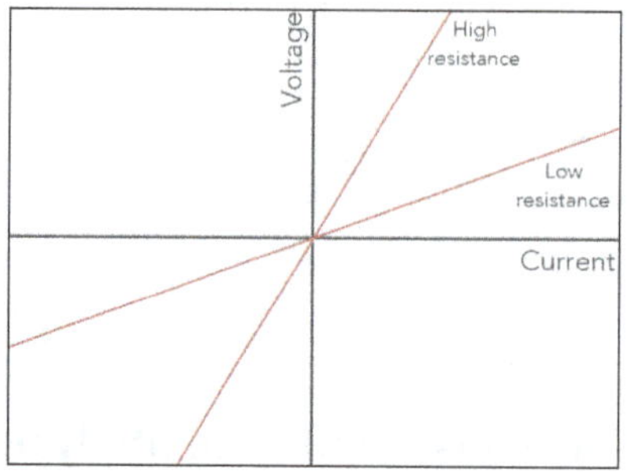

Plot of voltage and current for a linear resistance.

It can be seen that doubling the voltage doubles the current that passes through the particular circuit element.

On the graph, there are two lines, one for a higher resistance - this one requires more voltage to be applied for a given current flowing. Accordingly this must have a higher resistance. Conversely the curve for the lower resistance shows a component that requires a lower voltage to be applied for a given current.

Non-linear Components

Ohm's law in its basic form where a doubling in voltage results in a doubling in current applies to linear components like ordinary resistors. Some components like diodes have non-linear curves where the resistance is affected by the applied voltage.

Ohm's Law is one of the most basic concepts within electrical and electronic engineering. The concept of an item having a certain resistance which determines the amount of current that flows through it for a certain voltage is key to the operation virtually all circuits.

Ohm's Law Pie Chart

To help us understand the the relationship between the various values a little further, we can take all of the Ohm's Law equations from above for finding Voltage, Current, Resistance and of course Power and condense them into a simple Ohm's Law pie chart for use in AC and DC circuits and calculations as shown.

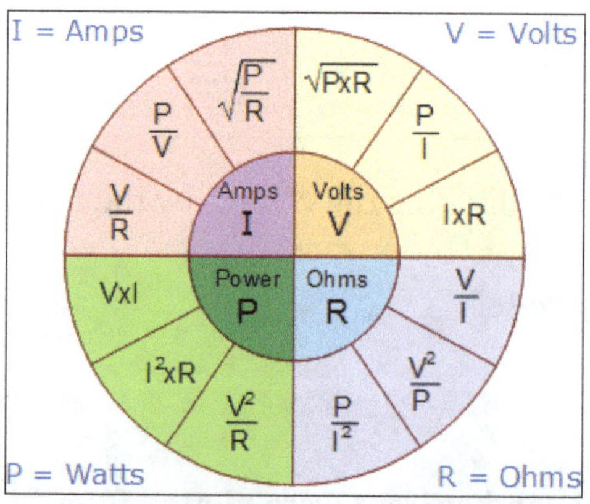

As well as using the Ohm's Law Pie Chart shown above, we can also put the individual Ohm's Law equations into a simple matrix table as shown for easy reference when calculating an unknown value.

Ohm's Law Matrix Table

Ohms Law Formulas				
Known Values	Resistance (R)	Current (I)	Voltage (V)	Power (P)
Current & Resistance	---	---	$V = I \times R$	$P = I^2 \times R$
Voltage & Current	$R = \frac{V}{I}$	---	---	$P = V \times I$
Power & Current	$R = \frac{P}{I^2}$	---	$V = \frac{P}{I}$	---
Voltage & Resistance	---	$I = \frac{V}{R}$	---	$P = \frac{V^2}{R}$
Power & Resistance	---	$I = \sqrt{\frac{P}{R}}$	$V = \sqrt{P \times R}$	---
Voltage & Power	$R = \frac{V^2}{P}$	$I = \frac{P}{V}$	---	---

Ohm's Law Example

For the circuit shown below find the Voltage (V), the Current (I), the Resistance (R) and the Power (P).

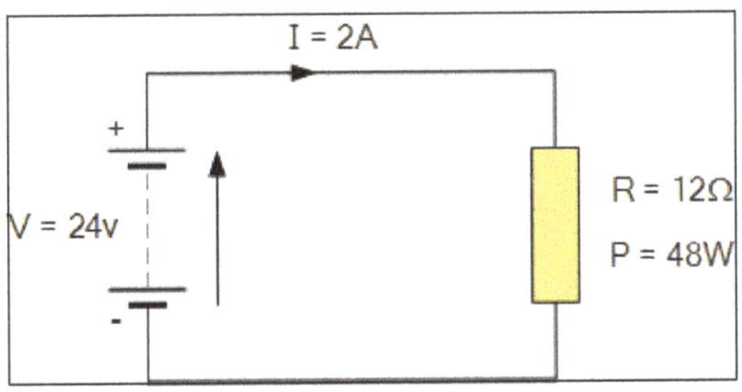

Voltage [V = I x R] = 2 x 12Ω = 24V

Current [I = V ÷ R] = 24 ÷ 12Ω = 2A

Resistance [R = V ÷ I] = 24 ÷ 2 = 12 Ω

Power [P = V x I] = 24 x 2 = 48W

Power within an electrical circuit is only present when BOTH voltage and current are present. For example, in an open-circuit condition, voltage is present but there is no current flow I = 0 (zero), therefore V*0 is 0 so the power dissipated within the circuit must also be 0. Likewise, if we have a short-circuit condition, current flow is present but there is no voltage V = 0, therefore 0*I = 0 so again the power dissipated within the circuit is 0.

As electrical power is the product of V*I, the power dissipated in a circuit is the same whether the circuit contains high voltage and low current or low voltage and high current flow. Generally, electrical power is dissipated in the form of Heat (heaters), Mechanical Work such as motors, Energy in the form of radiated (Lamps) or as stored energy (Batteries).

Kirchhoff's Circuit Laws

Kirchhoff's circuit laws are two equations first published by Gustav Kirchhoff in 1845. Fundamentally, they address conservation of energy and charge in the context of electrical circuits.

Although Kirchhoff's Laws can be derived from the equations of James Clerk Maxwell, Maxwell did not publish his set of differential equations (which form the foundation of classical electrodynamics, optics, and electric circuits) until 1861 and 1862. Kirchhoff, rather, used Georg Ohm's work as a foundation for *Kirchhoff's current law (KCL)* and *Kirchhoff's voltage law (KVL).*

Kirchhoff's laws are extremely important to the analysis of closed circuits. Consider, for example, the circuit illustrated in the figure below, consisting of five resistors in a combination of in series and parallel arrangements. Simplification of this circuit to a combination of series and parallel connections is impossible. However, using Kirchhoff's rules, one can analyze the circuit to determine the parameters of this circuit using the values of the resistors (R_1, R_2, R_3, r_1 and r_2). Also of importance in this example is that the values $E_1$and E_2 represent sources of voltage (e.g., batteries).

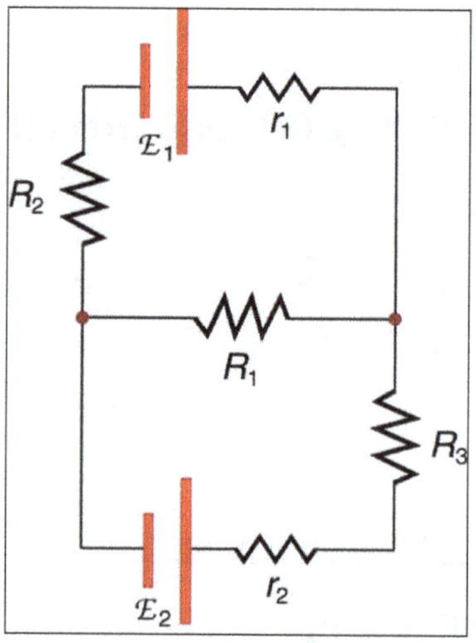

Closed Circuit: To determine all variables (i.e., current and voltage drops across the different resistors) in this circuit, Kirchhoff's rules must be applied.

As a final note, Kirchhoff's laws depend on certain conditions. The voltage law is a simplification of Faraday's law of induction, and is based on the assumption that there is no fluctuating magnetic field within the closed loop. Thus, although this law can be applied to circuits containing resistors and capacitors (as well as other circuit elements), it can only be used as an approximation to the behavior of the circuit when a changing current and therefore magnetic field are involved.

The Junction Rule

Kirchhoff's junction rule states that at any circuit junction, the sum of the currents flowing into and out of that junction are equal.

Kirchhoff's junction rule, also known as Kirchhoff's current law (KCL), Kirchoff's first law, Kirchhoff's point rule, and Kirchhoff's nodal rule, is an application of the principle of conservation of electric charge.

Kirchhoff's junction rule states that at any junction (node) in an electrical circuit, the sum of the currents flowing into that junction is equal to the sum of the currents flowing out of that junction. In other words, given that a current will be positive or negative depending on whether it is flowing towards or away from a junction, the algebraic sum of currents in a network of conductors meeting at a point is equal to zero.

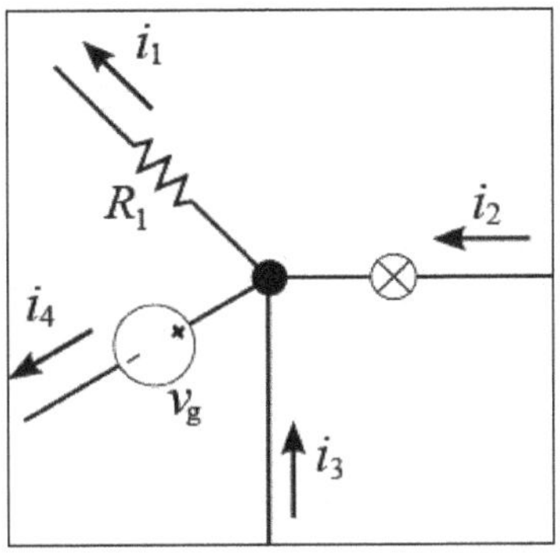

Kirchhoff's Junction Law illustrated as currents flowing into and out of a junction.

Thus, Kirchoff's junction rule can be stated mathematically as a sum of currents (I):

$$\sum_{k=1}^{n} I_k = 0$$

where n is the total number of branches carrying current towards or away from the node.

This law is founded on the conservation of charge (measured in coulombs), which is the product of current (amperes) and time (seconds).

Limitation

Kirchhoff's junction law is limited in its applicability. It holds for all cases in which total electric charge (Q) is constant in the region in consideration. Practically, this is always true so long as the law is applied for a specific point. Over a region, however, charge density may not be constant. Because charge is conserved, the only way this is possible is if there is a flow of charge across the boundary of the region. This flow would be a current, thus violating Kirchhoff's junction law.

The Loop Rule

Kirchhoff's loop rule states that the sum of the emf values in any closed loop is equal to the sum of the potential drops in that loop.

Kirchhoff's loop rule (otherwise known as Kirchhoff's voltage law (KVL), Kirchhoff's mesh rule, Kirchhoff's second law, or Kirchhoff's second rule) is a rule pertaining to circuits, and is based on the principle of conservation of energy.

Conservation of energy—the principle that energy is neither created nor destroyed—is a ubiquitous principle across many studies in physics, including circuits. Applied to circuitry, it is implicit that the directed sum of the electrical potential differences (voltages) around any closed network is equal to zero. In other words, the sum of the electromotive force (emf) values in any closed loop is equal to the sum of the potential drops in that loop (which may come from resistors).

Another equivalent statement is that the algebraic sum of the products of resistances of conductors (and currents in them) in a closed loop is equal to the total electromotive force available in that loop. Mathematically, Kirchhoff's loop rule can be represented as the sum of voltages in a circuit, which is equated with zero:

$$\sum_{k=1}^{n} V_k = 0.$$

Here, V_k is the voltage across element k, and n is the total number of elements in the closed loop circuit. An illustration of such a circuit is shown in figure. In this example, the sum of v_1, v_2, v_3, and v_4 (and v_5 if it is included), is zero.

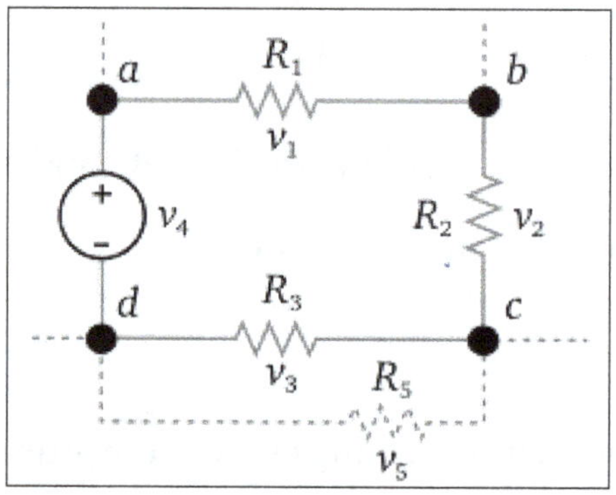

Kirchhoff's Loop Rule: Kirchhoff's loop rule states that the sum of all the voltages around the loop is equal to zero: v1 + v2 + v3 – v4 = 0.

Given that voltage is a measurement of energy per unit charge, Kirchhoff's loop rule is based on the law of conservation of energy, which states: the total energy gained per unit charge must equal the amount of energy lost per unit of charge.

Example

The following illustrates the changes in potential in a simple series circuit loop. Kirchhoff's second rule requires emf–Ir–IR_1–IR_2=0. Rearranged, this is emf=Ir+IR_1+IR_2, meaning that the emf equals the sum of the IR (voltage) drops in the loop. The emf supplies 18 V, which is reduced to zero by the resistances, with 1 V across the internal resistance, and 12 V and 5 V across the two load resistances, for a total of 18 V.

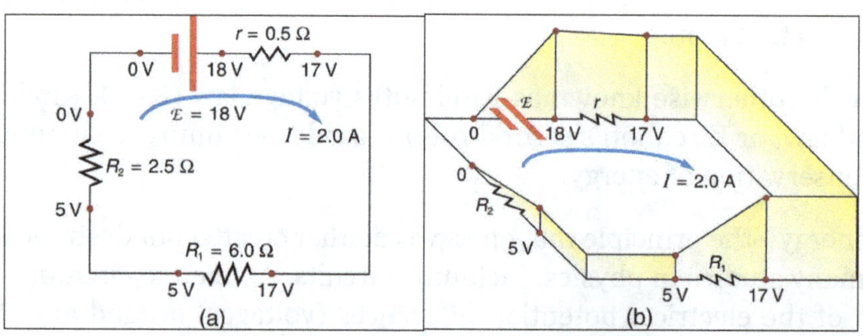

The Loop Rule: An example of Kirchhoff's second rule where the sum of the changes in potential around a closed loop must be zero.

(a) In this standard schematic of a simple series circuit, the emf supplies 18 V, which is reduced to zero by the resistances, with 1 V across the internal resistance, and 12 V and 5 V across the two load resistances, for a total of 18 V.

(b) This perspective view represents the potential as something like a roller coaster, where charge is raised in potential by the emf and lowered by the resistances.

Limitation

Kirchhoff's loop rule is a simplification of Faraday's law of induction, and holds under the assumption that there is no fluctuating magnetic field linking the closed loop. In the presence of a variable magnetic field, electric fields could be induced and emf could be produced, in which case Kirchhoff's loop rule breaks down.

Applications

Kirchhoff's rules can be used to analyze any circuit and modified for those with EMFs, resistors, capacitors and more. Kirchhoff's rules can be used to analyze any circuit by modifying them for those circuits with electromotive forces, resistors, capacitors and more. Practically speaking, however, the rules are only useful for characterizing those circuits that cannot be simplified by combining elements in series and parallel.

Combinations in series and parallel are typically much easier to perform than applying either of Kirchhoff's rules, but Kirchhoff's rules are more broadly applicable and should be used to solve problems involving complex circuits that cannot be simplified by combining circuit elements in series or parallel.

Example of Kirchoff's Rules

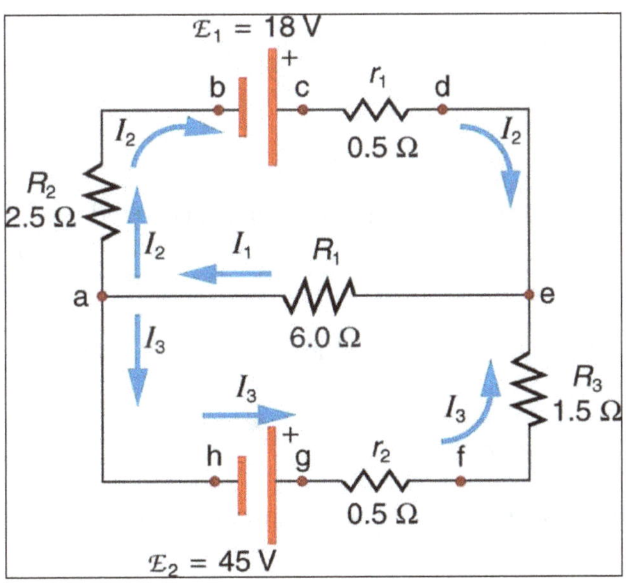

Shows a very complex circuit, but Kirchhoff's loop and junction rules can be applied. To solve the circuit for currents I_1, I_2, and I_3, both rules are necessary:

Kirchhoff's Rules sample problem: This figure shows a very complicated circuit, which can be reduced and solved using Kirchoff's Rules.

Applying Kirchhoff's junction rule at point a, we find:

$$I_1 = I_2 + I_3$$

because I_1 flows into point a, while I_2 and I_3 flow out. The same can be found at point e. We now must solve this equation for each of the three unknown variables, which will require three different equations.

Considering loop abcdea, we can use Kirchhoff's loop rule:

$$-I_2R_2 + emf_1 - I_2r_1 - I_1R_1 = -I_2(R_2) + r_1) + emf_1 - I_1R_1 = 0$$

Substituting values of resistance and emf from the figure diagram and canceling the ampere unit gives:

$$-3I_2 + 18 - 6I_1 = 0$$

This is the second part of a system of three equations that we can use to find all three current values. The last can be found by applying the loop rule to loop aefgha, which gives:

$$I_1R_1 + I_3R_3 + I_3r_2 - emf_2 = I_1R_1 + I_3(R_3 + r_2) - emf_2 = 0$$

Using substitution and simplifying, this becomes:

$$6I_1 + 2I_3 - 45 = 0$$

In this case, the signs were reversed compared with the other loop, because elements are traversed in the opposite direction.

We now have three equations that can be used in a system. The second will be used to define I_2, and can be rearranged to:

$$I_2 = 6 - 2I_1$$

The third equation can be used to define I_3, and can be rearranged to:

$$I_3 = 22.5 - 3I_1$$

Substituting the new definitions of I_2 and I_3 (which are both in common terms of I_1), into the first equation ($I_1=I_2+I_3$), we get:

$$I_1 = (6 - 2I_1) + (22.5 - 3I_1) = 28.5 - 5I_1$$

Simplifying, we find that I_1=4.75 A. Inserting this value into the other two equations, we find that I_2=-3.50 A and I_3=8.25 A.

Source Transformation

Source Transformation simply means replacing one source by an equivalent source. A practical voltage source can be transformed into an equivalent practical current source and similarly a practical current source into voltage source.

Any practical voltage source or simply a voltage source consists of an ideal voltage source in series with an internal resistance or impedance (for any ideal source this impedance will be zero), the output voltage becomes independent of the load current. Cells, batteries and generators are the example of the voltage source.

For any practical current source or simply current source, is an ideal current source in parallel with the internal resistance or impedance, for ideal current source this parallel impedance is infinity. The semiconductor devices like transistors, etc are treated as a current source or an output produce by direct or alternating voltage source is called as direct and alternating current source, respectively.

The voltage and current source are mutually transferable or in other words the source transformation i.e. voltage to current source and current to voltage source can be done. Let us understand this by considering a circuit given below:

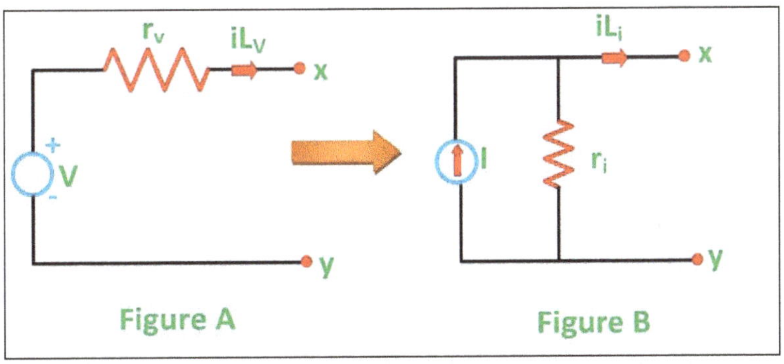

Figure A represents a practical voltage source in series with the internal resistance r_v, while figure B represents a practical current source with parallel internal resistance r_i.

For the practical voltage source the load current will be given by the equation,

$$iL_v = \frac{V}{r_V + r_L}$$

Where,

iL_v is the load current for the practical voltage source.

V is the voltage.

r_v is the internal resistance of the voltage source.

r_L is the load resistance.

It is assumed that the load resistance r_L is connected at the terminal x-y. Similarly for the practical current source the load current is given as:

$$iL_i = I\frac{r_i}{r_i + r_L}$$

Where,

iL_i is the load current for the practical current source.

I is the current.

r_i is the internal resistance of the current source.

r_L is the load resistance connected across the terminal x-y in the figure B.

Two sources become identical, when we will equate the above two equation,

$$\frac{V}{r_V + r_L} = I\frac{r_i}{r_i + r_L}$$

However, for the current source the terminal voltage at x-y would be Iri, x-y terminal are open. i.e.,

$$V = I \times r_i$$

Therefore, we will get,

$$r_V + r_L = r_i + r_L \text{ or}$$
$$r_V = r_L$$

Therefore, for any practical voltage source, if the ideal voltage be V and internal resistance be r_v, the voltage source can be replaced by a current source I with the internal resistance in parallel with the current source.

Source Transformation : Conversion of Voltage Source into Current Source.

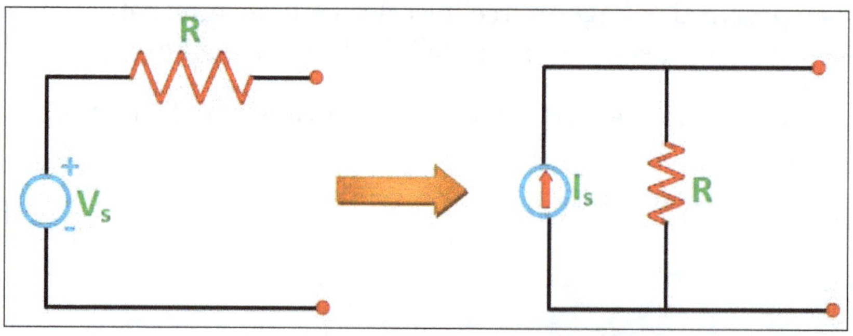

When the voltage source is connected with the resistance in series and it has to be converted into the current source than the resistance is connected in parallel with the current source as shown in the above figure.

Where, $I_s = V_s/R$

Conversion of Current Source into Voltage Source

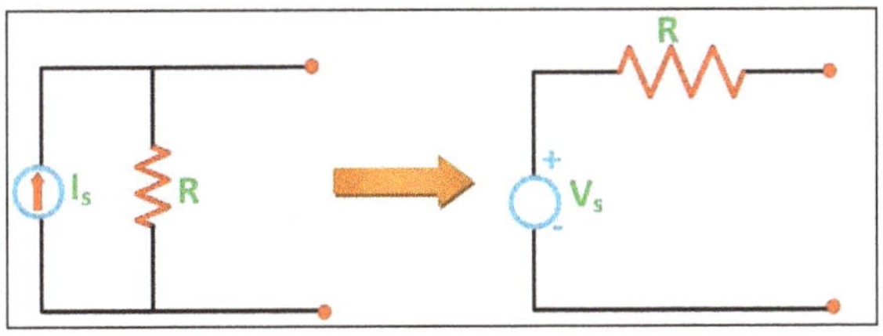

In the above circuit diagram a current source which is connected in parallel with the resistance is transformed into a voltage source by placing the resistance in series with the voltage source.

Where, $V_s = I_s / R$.

Substitution Theorem

Substitution Theorem states that the voltage across any branch or the current through that branch of a network being known, the branch can be replaced by the combination of various elements that will make the same voltage and current through that branch. In other words, the Substitution Theorem says that for branch equivalence, the terminal voltage and current must be same.

The concept of the theorem is based on the substitution of one element from another element. This theorem gives intuition on the behaviour of the circuit. It also helps in proving several other theorems. But the substitution theorem cannot use for solving the theorem which has more than two sources which are neither connected in series nor parallel.

Explanation of Substitution Theorem

Simply one can say that the Substitution Theorem is the replacement of one element with another equivalent element. In a network, if any element is substituted or replaced by a voltage or current source whose voltage and current across or through that element will remain unchanged as the previous network.

Let us understand the theorem with the help of the circuit diagram shown below:

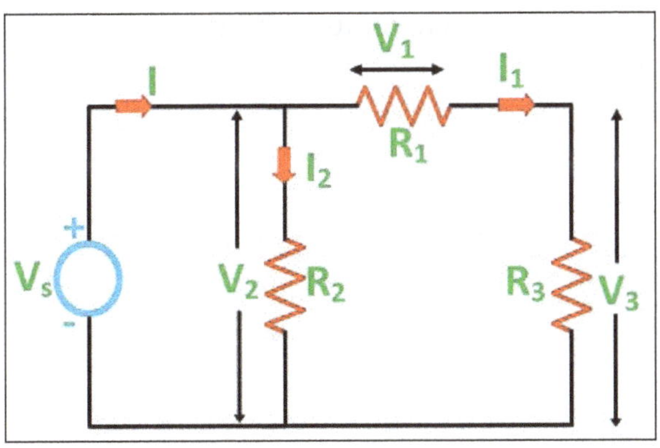

Here various resistances R_1, R_2, R_3 is connected across with the voltage source (V). Current I is flowing through the circuit, which is divided into current I_1 flowing through the resistance R_1 and the current I_2 flowing through the resistance R_2. V_1, V_2 and V_3 are the voltage drop across the resistance R_1, R_2 and R_3 respectively.

Now if the resistance R_3 is substituted by the voltage source V_3 as shown in the circuit diagram below,

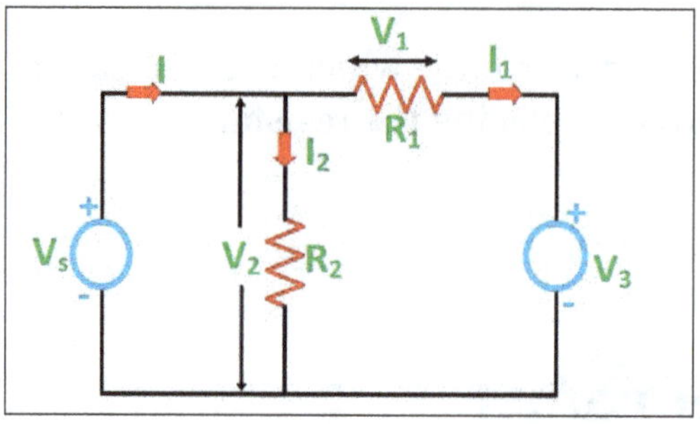

In the circuit diagram shown below the resistance, R_3 is replaced by the current flowing through that element, i.e. I_1.

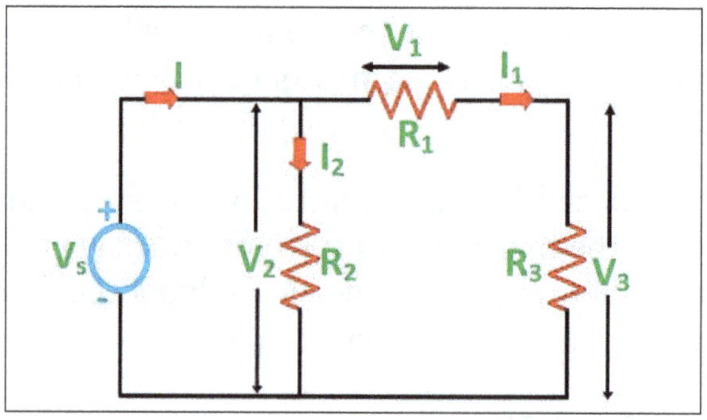

In both the cases shown above if the element is substituted by the voltage source or the current source, then also, the initial conditions of the circuit does not alter. This means that the voltage across the resistance and current flowing through the resistance unaltered even if they are substituted by other sources.

Steps for Solving Network Using Substitution Theorem

Step 1 – First obtain the concerned branch voltage and current flowing through the branch given by V_{xy} and I_{xy} as shown below in the figure A.

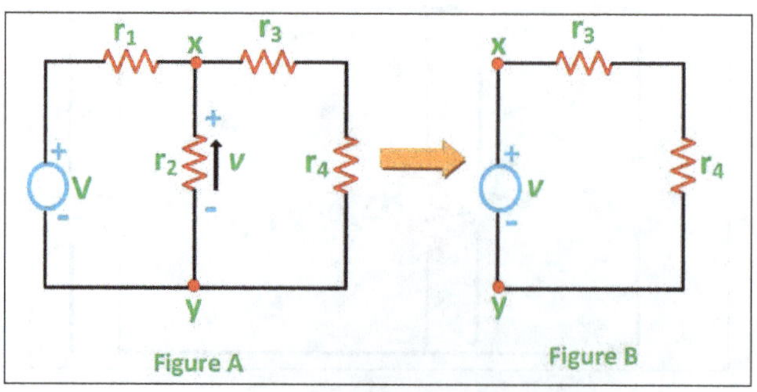

Step 2 – The branch may be substituted by an independent voltage source as shown above in figure B.

Step 3 – Similarly the branch may be replaced by an independent current source as shown below in the circuit diagram C.

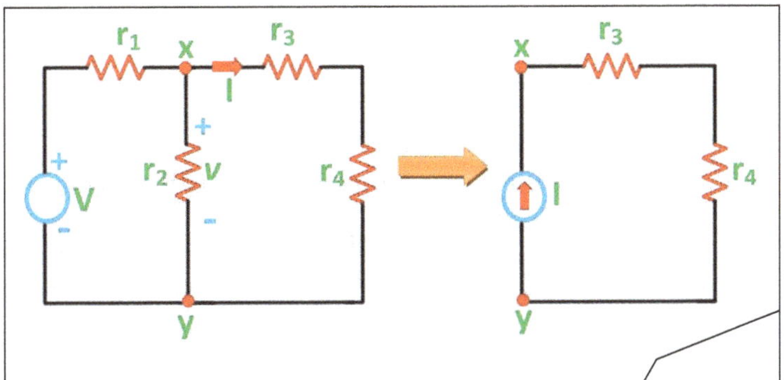

Figure C

Step 4 – Hence it is seen that the voltage drop and the current flowing through the circuit A shown above will be same if it is substituted by any independent voltage or current source shown in the figure B and C.

Norton's Theorem

Norton's theorem states that any linear complex electrical circuit can be reduced into a simple electric circuit with one current and resistance connected in parallel. For understanding in depth regarding norton theory, let us consider Norton's theorem examples as follows:

Examples

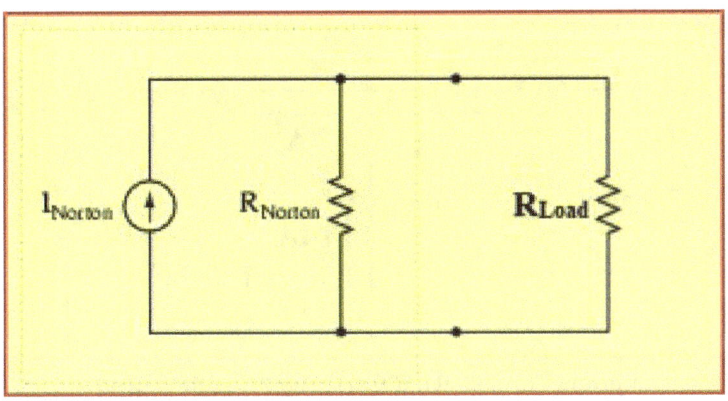

Primarily, let us consider a simple electrical circuit that consists of two voltage sources and three resistors which are connected as shown in the above figure. The above circuit consists of three resistors among which R_2 resistor is considered as load. Then, the circuit can be represented as shown below.

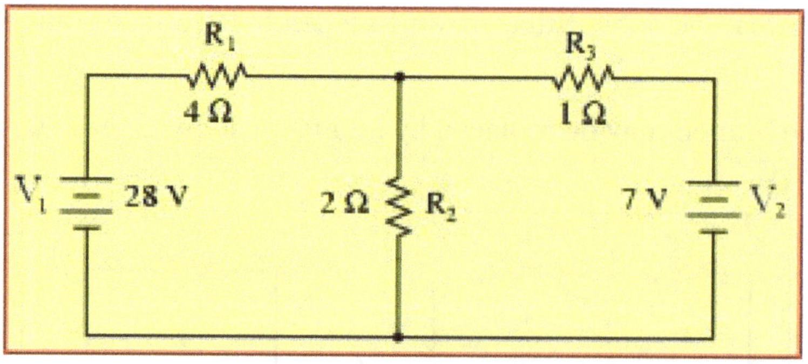

Nortons Theorem Example Circuit with Load Resistor.

We know that, if the load changes, then the calculation of various parameters of electric circuits is difficult. So, network theorems are used for calculating the network parameters easily.

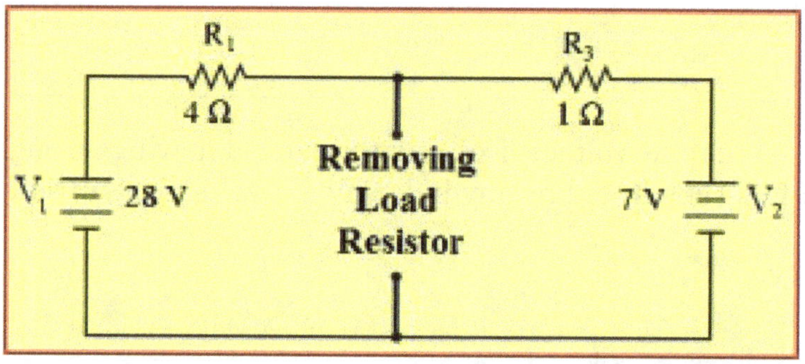

Nortons Theorem Example Circuit After Removing Load Resistor.

In this Norton's theorem also we follow the procedure similar to thevenins theorem (up to some extent). Here, primarily remove the load (consider resistor R_2=2 Ohms as load in the circuit) as shown in the above figure. Then, short circuit the load terminals with a wire (exactly opposite to the procedure that we follow in thevenins theorem, i.e., open circuit of load terminals) as shown in the below figure. Now, calculate the resultant current (current through resistors R_1, R_3, and short circuit line after removing R_2) as shown in the figure below:

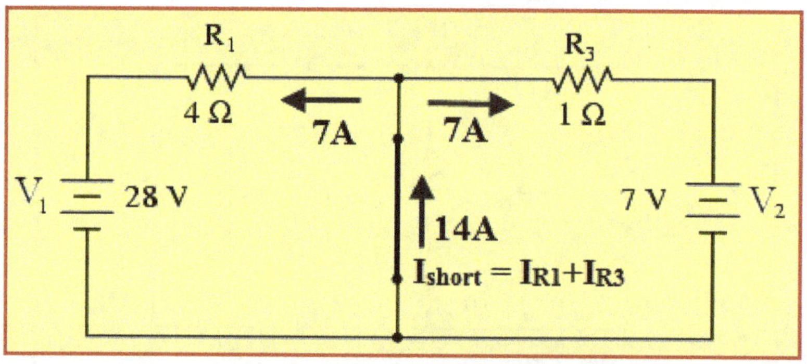

Current Through R1, R3, and Short Circuited Load.

From the above figure, the Nortons source current is equal to 14A which is used in the Norton's equivalent circuit as shown in the below figure. Norton's theorem equivalent circuit consists of the Norton current source (INorton) in parallel with Norton's equivalent resistance (RNorton) and load (here R_2=2Ohms).

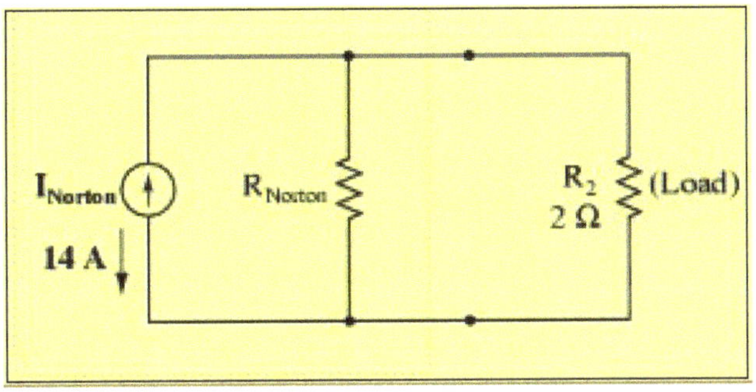

Nortons Equivalent Circuit with INorton, RNorton, RLoad.

This Nortorn's theorem equivalent circuit is a simple parallel circuit as shown in the figure. Now, for calculating Norton's equivalent resistance we have to follow two procedures such as Thevenins theorem and Superposition theorem.

Primarily, remove the load resistance (similar to thevenins theorem step of calculating thevenins resistance). Then, replace voltage sources with short circuit (wires in case of ideal voltage sources and in case of practical voltage sources their internal resistances are used). Similarly, current sources with open circuit (breaks in case of ideal current sources and in case of practical current sources their internal resistances are used). Now, the circuit becomes as shown in the figure below and it is a simple parallel circuit with resistors.

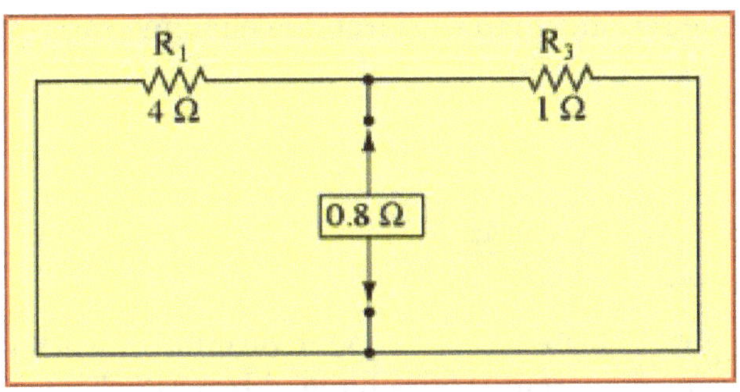

Finding Nortons Resistance.

As the resistors R_1 and R_3 are parallel to each other, the value of Norton's resistance is equal to parallel resistance value of R_1 and R_3. Then, the total Norton's theorem equivalent circuit can be represented as shown in the circuit below:

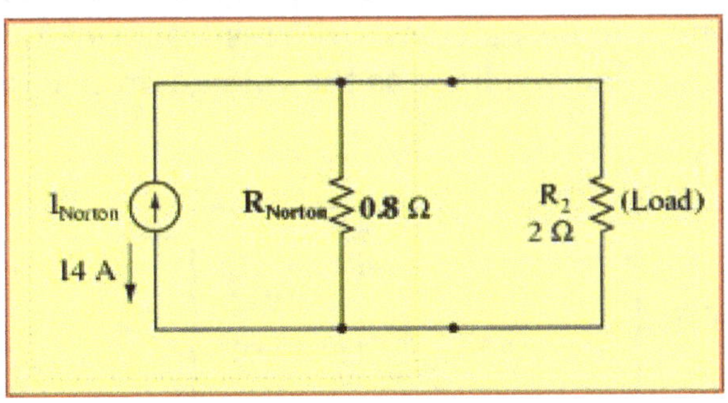

Norton's Theorem Equivalent Circuit.

The formula for calculating the load current, Iload can be calculated using various basic laws such as Ohm's law, Krichhoff's voltage law, and Krichhoff's current law.

Thus, the current passing through the load resistor R load (R_2) is given by,

$$I_L = \frac{I_N . R_N}{R_N + R_L}$$

Load Current formula.

Where,

I_N = Norton's current (14A)

R_N = Norton's resistance (0.8 Ohms)

R_L = Load resistance (2 Ohms)

Therefore, I load = current passing through load resistance = 4A.

Similarly, the large, complex, linear networks with several numbers of sources (current or voltage sources) and resistors can be reduced to simple parallel circuits with single current source in parallel with Norton's resistance and load.

Thus, the Norton's equivalent circuit with Rn and In can be determined and a simple parallel circuit can be formed (from a complex network circuit). The calculations of the circuit parameters can be easily analyzed. If one resistance in the circuit is changed rapidly (load), then the Norton's theorem can be used to perform calculations easily.

Compensation Theorem

Compensation Theorem states that in a linear time invariant network when the resistance (R) of an uncoupled branch, carrying a current (I), is changed by (ΔR). The currents in all the branches would change and can be obtained by assuming that an ideal voltage source of (V_C) has been connected such that V_C= I (ΔR) in series with (R + ΔR) when all other sources in the network are replaced by their internal resistances.

In Compensation Theorem, the source voltage (V_C) opposes the original current. In simple words compensation theorem can be stated as – the resistance of any network can be replaced by a voltage source, having the same voltage as the voltage drop across the resistance which is replaced.

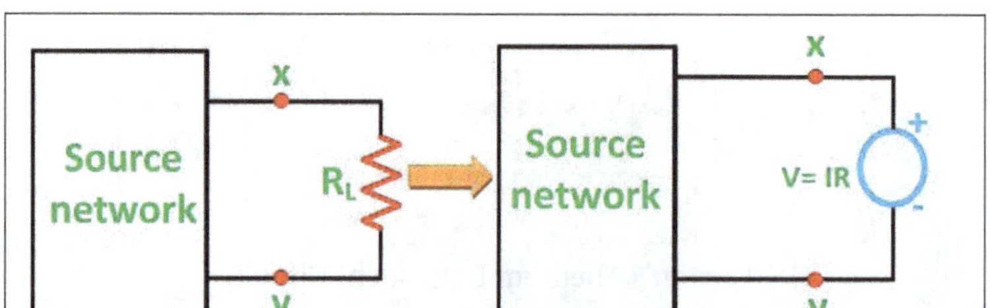

Let us assume a load R_L be connected to a DC source network whose Thevenin's equivalent gives V_o as the Thevenin's voltage and R_{TH} as the Thevenin's resistance as shown in the figure below:

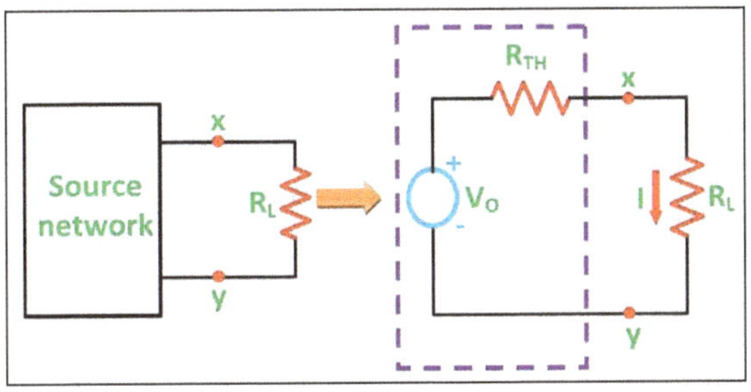

Here,

$$I = \frac{V_0}{R_{TH} + R_L}$$

Let the load resistance R_L be changed to $(R_L + \Delta R_L)$. Since the rest of the circuit remains unchanged, the Thevenin's equivalent network remains the same as shown in the circuit diagram below:

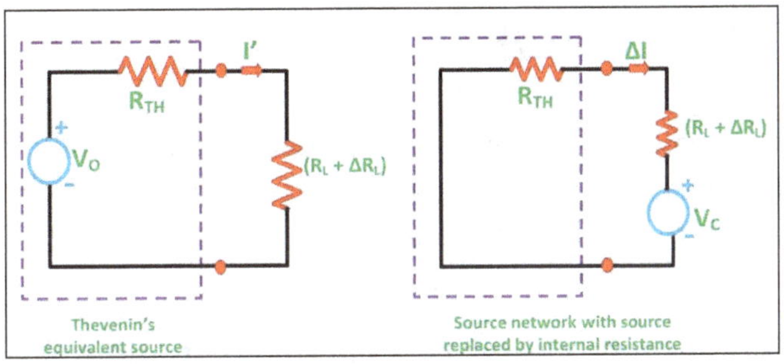

Here,

$$I' = \frac{V_0}{R_{TH} + (R_L + \Delta R_L)}$$

The change of current being termed as ΔI,

Therefore,

$$\Delta I = I' - I$$

Putting the value of I' and I from the above three equation we will get the following equation,

$$\Delta I = \frac{V_0}{R_{TH} + (R_L + \Delta R_L)} - \frac{V_0}{R_{TH} + R_L}$$

$$\Delta I = \frac{V_0\{R_{TH} + R_L - (R_{TH} + R_L + \Delta R_L)\}}{(R_{TH} + R_L + \Delta R_L)(R_{TH} + R_L)}$$

$$\Delta I = -\left[\frac{V_0}{R_{TH} + R_L}\right]\frac{\Delta R_L}{R_{TH} + R_L + \Delta R_L}$$

Now, putting the value of I from the equation mentioned in the starting of this topic in the equation above, we will get the following equation,

$$\Delta I = -\frac{I\Delta R_L}{R_{TH} + R_L + \Delta R_L}$$

As we know, $V_C = I\,\Delta\,R_L$ and is known as compensating voltage.

Therefore, the above equation becomes,

$$\Delta I = -\frac{-V_C}{R_{TH} + R_L + \Delta R_L}$$

Hence, Compensation Theorem tells that with the change of branch resistance, branch currents changes and the change is equivalent to an ideal compensating voltage source in series with the branch opposing the original current, all other sources in the network being replaced by their internal resistances.

Superposition Theorem

Superposition theorem is based on the concept of linearity between the response and excitation of an electrical circuit. It states that the response in a particular branch of a linear circuit when multiple independent sources are acting at the same time is equivalent to the sum of the responses due to each independent source acting at a time.

In this method, we will consider only one independent source at a time. So, we have to eliminate the remaining independent sources from the circuit. We can eliminate the voltage sources by shorting their two terminals and similarly, the current sources by opening their two terminals.

Therefore, we need to find the response in a particular branch 'n' times if there are 'n' independent sources. The response in a particular branch could be either current flowing through that branch or voltage across that branch.

Procedure of Superposition Theorem

Follow these steps in order to find the response in a particular branch using superposition theorem.

- Step 1: Find the response in a particular branch by considering one independent source and eliminating the remaining independent sources present in the network.
- Step 2: Repeat Step 1 for all independent sources present in the network.
- Step 3: Add all the responses in order to get the overall response in a particular branch when all independent sources are present in the network.

Example: Find the current flowing through 20 Ω resistor of the following circuit using superposition theorem.

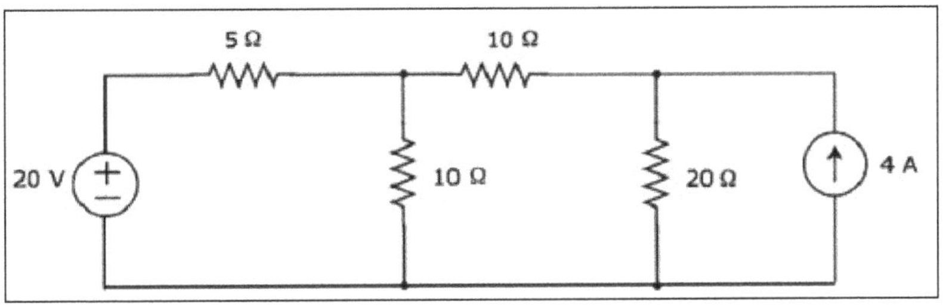

Step 1

Let us find the current flowing through 20 Ω resistor by considering only 20 V voltage source. In this case, we can eliminate the 4 A current source by making open circuit of it. The modified circuit diagram is shown in the following figure.

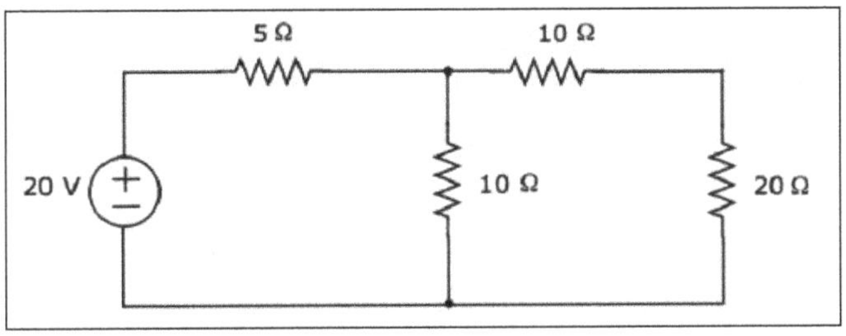

There is only one principal node except Ground in the above circuit. So, we can use nodal analysis method. The node voltage V_1 is labelled in the following figure. Here, V_1 is the voltage from node 1 with respect to ground.

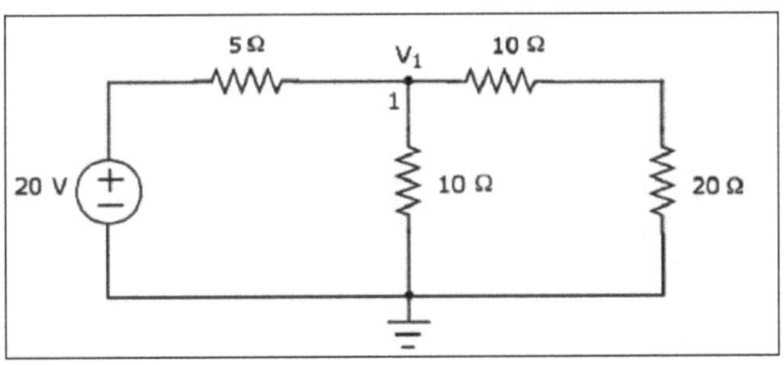

The nodal equation at node 1 is,

$$\frac{V_1 - 20}{5} + \frac{V_1}{10} + \frac{V_1}{10+20} = 0$$

$$\Rightarrow \frac{6V_1 - 120 + 3V_1 + V_1}{30} = 0$$

$$\Rightarrow 10V_1 = 120$$

$$\Rightarrow V_1 = 12\,V$$

The current flowing through 20 Ω resistor can be found by doing the following simplification.

$$I_1 = \frac{V_1}{10+20}$$

Substitute the value of V_1 in the above equation.

$$I_1 = \frac{12}{10+20} = \frac{12}{30} = 0.4\,A$$

Therefore, the current flowing through 20 Ω resistor is 0.4 A, when only 20 V voltage source is considered.

Step 2

Let us find the current flowing through 20 Ω resistor by considering only 4 A current source. In this case, we can eliminate the 20 V voltage source by making short-circuit of it. The modified circuit diagram is shown in the following figure.

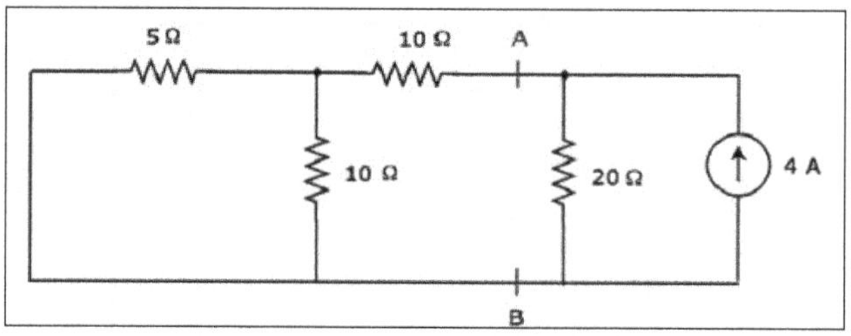

$$R_{AB} = \left(\frac{5\times10}{5+10}\right) + 10 = \frac{10}{3} + 10 = \frac{40}{3}\Omega$$

The simplified circuit diagram is shown in the following figure.

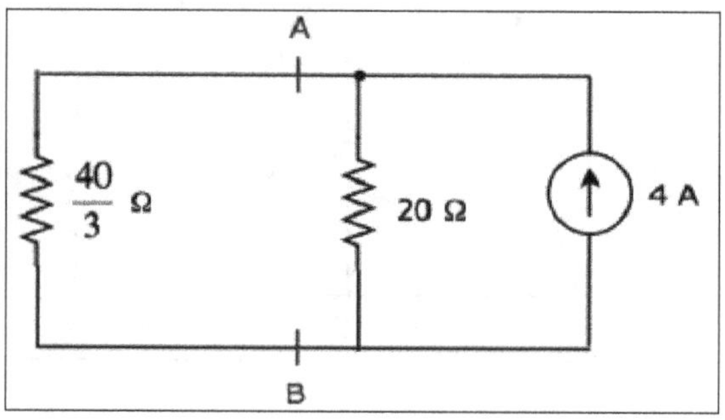

We can find the current flowing through 20 Ω resistor, by using current division principle.

$$I_2 = I_S\left(\frac{R_1}{R_1+R_2}\right)$$

Substitute I_S=4A, $R_1=\frac{40}{3}$ Ω and R_2=20Ω in the above equation.

$$I_2 = 4\left(\frac{\frac{40}{3}}{\frac{40}{3}+20}\right) = 4\left(\frac{40}{100}\right) = 1.6A$$

Therefore, the current flowing through 20 Ω resistor is 1.6 A, when only 4 A current source is considered.

Step 3

We will get the current flowing through 20 Ω resistor of the given circuit by doing the addition of two currents that we got in step 1 and step 2. Mathematically, it can be written as

$$I = I_1 + I_2$$

Substitute, the values of I_1 and I_2 in the above equation.

$$I = 0.4 + 1.6 = 2A$$

Therefore, the current flowing through 20 Ω resistor of given circuit is 2A.

Thevenin's Theorem

Thevenin's Theorem states that – any complicated network across its load terminals can be substituted by a voltage source with one resistance in series. This theorem helps in the study of the variation of current in a particular branch when the resistance of the branch is varied while the remaining network remains the same for example designing of electronics circuits.

A more general statement of Thevenin's Theorem is that any linear active network consisting of independent or dependent voltage and current source and the network elements can be replaced by an equivalent circuit having a voltage source in series with a resistance, that voltage source being the open circuited voltage across the open circuited load terminals and the resistance being the internal resistance of the source.

In other words, the current flowing through a resistor connected across any two terminals of a network by an equivalent circuit having a voltage source E_{th} in series with a resistor R_{th}. Where Eth is the open circuit voltage between the required two terminals called the Thevenin voltage and the R_{th} is the equivalent resistance of the network as seen from the two terminal with all other sources replaced by their internal resistances called Thevenin resistance.

Explanation of Thevenin's Theorem

The Thevenin's statement is explained with the help of a circuit shown below.

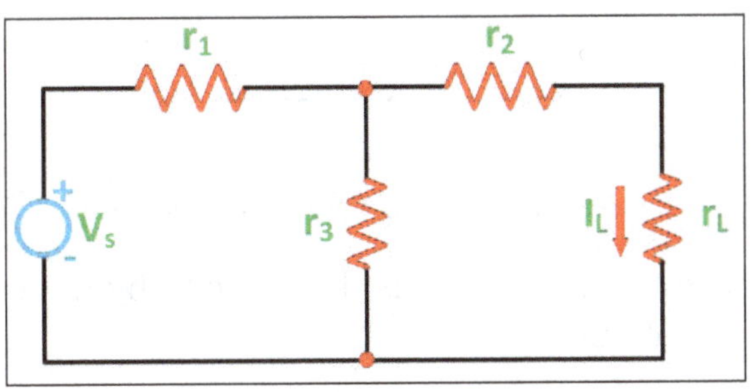

Let us consider a simple DC circuit as shown in the figure above, where we have to find the load current I_L by the Thevenin's theorem. In order to find the equivalent voltage source, r_L is removed from the circuit as shown in the figure below and V_{oc} or V_{TH} is calculated.

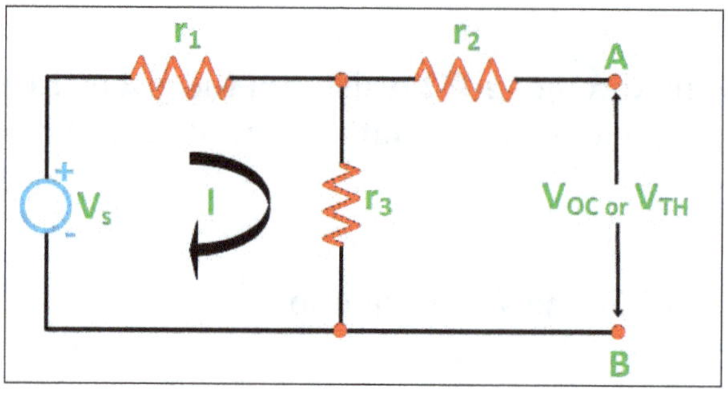

$$V_{OC} = I\, r_3 = \frac{V_S}{r_1 + r_3} r_3$$

Now, to find the internal resistance of the network (Thevenin's resistance or equivalent resistance) in series with the open circuit voltage V_{OC}, also known as Thevenin's voltage V_{TH}, the voltage source is removed or we can say it is deactivated by a short circuit (as the source does not have any internal resistance) as shown in the figure below:

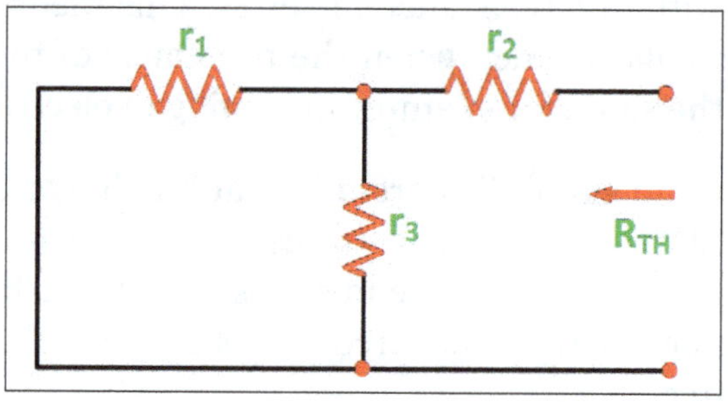

$$R_{TH} = r_2 + \frac{r_1 r_3}{r_1 + r_3}$$

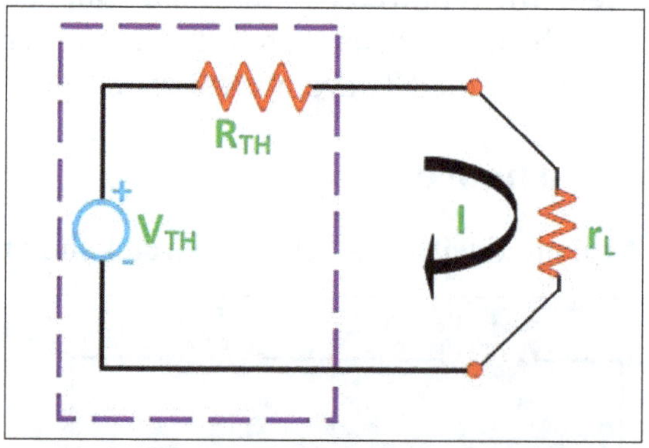

Equivalent Circuit of Thevenin's Theorem.

As per Thevenin's Statement, the load current is determined by the circuit shown above and the equivalent Thevenin's circuit is obtained.

The Load current I_L is given as,

$$I_L = \frac{V_{TH}}{R_{TH} + r_L}$$

Where,

V_{TH} is the Thevenin's equivalent voltage. It is an open circuit voltage across the terminal AB known as load terminal R_{TH} is the Thevenin's equivalent resistance, as seen from the load terminals where all the sources are replaced by their internal impedance r_L is the load resistance.

Steps for Solving Thevenin's Theorem

- Step 1: First of all remove the load resistance r_L of the given circuit.
- Step 2: Replace all the impedance source by their internal resistance.
- Step 3: If sources are ideal then short circuit the voltage source and open the current source.
- Step 4: Now find the equivalent resistance at the load terminals know as Thevenin's Resistance (R_{TH}).
- Step 5: Draw the Thevenin's equivalent circuit by connecting the load resistance and after that determine the desired response.

This theorem is possibly the most extensively used networks theorem. It is applicable where it is desired to determine the current through or voltage across any one element in a network. The Thevenin's Theorem is an easy way to solve the complicated network.

Example

For example, consider the following circuit;

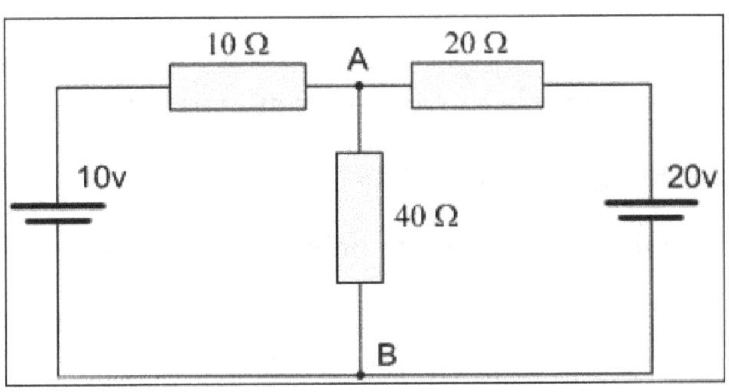

Firstly, we have to remove the centre 40Ω resistor and short out (not physically as this would be dangerous) all the emf´s connected to the circuit, or open circuit any current sources. The value of resistor Rs is found by calculating the total resistance at the terminals A and B with all the emf´s removed, and the value of the voltage required Vs is the total voltage across terminals A and B with an open circuit and no load resistor Rs connected. Then, we get the following circuit;

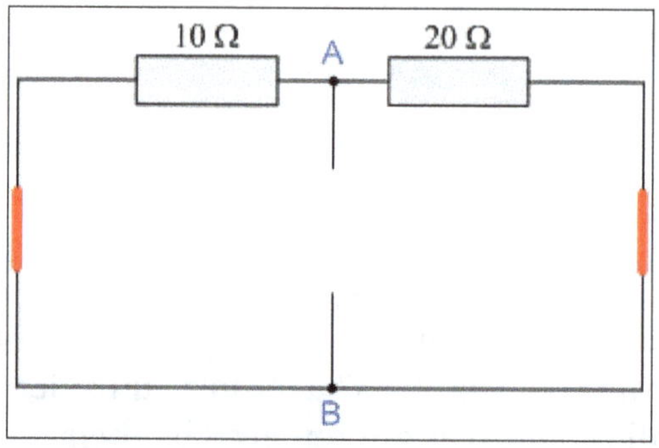

Find the Equivalent Resistance (Rs):

10ΩResistor in parallel with the 20ΩResistor,

$$R_T = \frac{R_1 \times R_2}{R_1 + R_2} = \frac{20 \times 10}{20 + 10} = 6.67\Omega$$

Find the Equivalent voltage (Vs):

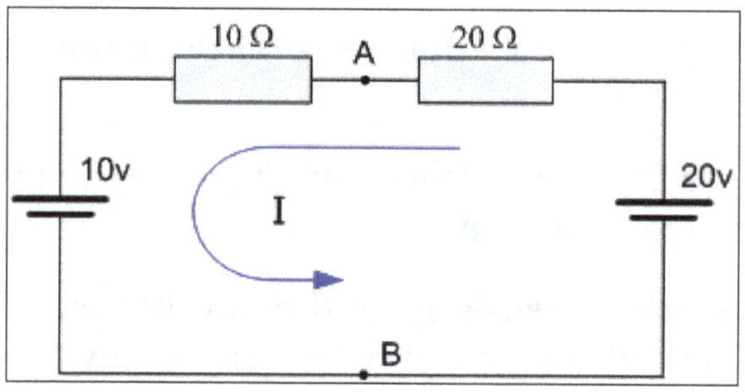

We now need to reconnect the two voltages back into the circuit, and as $V_S = V_{AB}$ the current flowing around the loop is calculated as:

$$I = \frac{20v - 10v}{20\Omega + 10\Omega} = 0.33amps$$

So the voltage drop across the 20Ω resistor can be calculated as:

V_{AB} = 20 - (20Ω x 0.33amps) = 13.33 volts.

Then the Thevenin's Equivalent circuit is shown below with the 40Ω resistor connected.

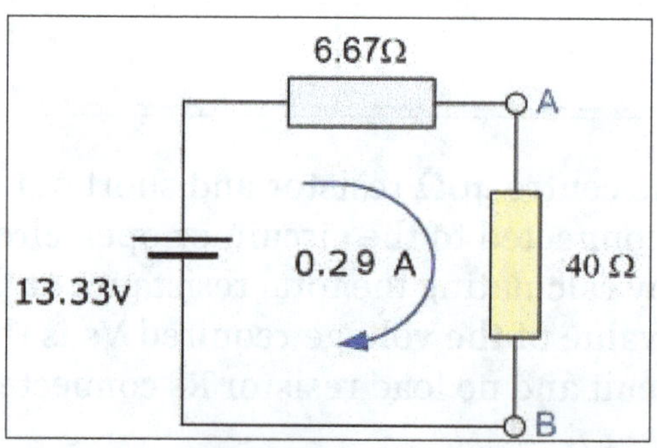

And from this the current flowing in the circuit is given as:

$$I = \frac{13.33v}{6.67\Omega + 40\Omega} = 0.29amps$$

Thevenin's theorem can be used as a circuit analysis method and is particularly useful if the load is to take a series of different values. It is not as powerful as Mesh or Nodal analysis in larger networks because the use of Mesh or Nodal analysis is usually necessary in any Thevenin exercise, so it might well be used from the start. However, Thevenin's equivalent circuits of Transistors, Voltage Sources such as batteries etc, are very useful in circuit design.

Applications of Thevenin's Theorem

- Thevenin's Theorem is especially useful in analyzing power systems and other circuits where one particular resistor in the circuit (called the "load" resistor) is subject to change, and re-calculation of the circuit is necessary with each trial value of load resistance, to determine voltage across it and current through it.
- Source modeling and resistance measurement using the Wheatstone bridge provide applications for Thevenin's theorem.

Practical Limitations

- Many, if not most circuits are only linear over a certain range of values, thus the Thévenin equivalent is valid only within this linear range and may not be valid outside the range.
- The Thévenin equivalent has an equivalent I-V characteristic only from the point of view of the load.
- The power dissipation of the Thévenin equivalent is not necessarily identical to the power dissipation of the real system. However, the power dissipated by an external resistor between the two output terminals is the same however the internal circuit is represented.

References

- Network-analysis-for-electric-circuits: electricalengineeringforbeginners.blogspot.com, Retrieved 10 April 2020
- Electric-circuit-or-electrical-network: electrical4u.com, Retrieved 28 May 2020
- What-is-rl-series-circuit: circuitglobe.com, Retrieved 18 May 2020
- What-is-an-rc-resistor-capacitor-circuit: circuitglobe.com, Retrieved 09 July 2020
- What-is-substitution-theorem: circuitglobe.com, Retrieved 08 March 2020
- Electrical-impedance-and-its-applications: elprocus.com, Retrieved 23 July 2020
- What-is-an-electrical-circuit, electricity, science: eschooltoday.com, Retrieved 15 April 2020
- Active-and-passive-elements-of-electrical-circuit: electrical4u.com, Retrieved 28 May 2020

5

Integrated Circuit

An assembly of electronic circuits on a thin substrate of semiconductor is known as an integrated circuit. The logic and techniques required to design integrated circuits fall under integrated circuit design. This chapter discusses in detail the theories and concepts related to integrated circuit ad integrated circuit design.

An integrated circuit or monolithic integrated circuit (also referred to as an IC, a chip, or a microchip) is a set of electronic circuits on one small plate ("chip") of semiconductor material, normally silicon. This can be made much smaller than a discrete circuit made from independent electronic components. ICs can be made very compact, having up to several billion transistors and other electronic components in an area the size of a human fingernail. The half-pitch between nodes in a circuit has been made smaller as the technology advances; in 2008 it dropped below 100 nanometers, and was reduced to around 14 nanometers in 2014.

Erasable programmable read-only memory integrated circuits. These packages have a transparent window that shows the die inside. The window allows the memory to be erased by exposing the chip to ultraviolet light.

ICs were made possible by experimental discoveries showing that semiconductor devices could perform the functions of vacuum tubes and by mid-20th-century technology advancements in semiconductor device fabrication. The integration of large numbers of tiny transistors into a small chip was an enormous improvement over the manual assembly of circuits using discrete electronic components. The integrated circuit's mass production capability, reliability and building-block approach to circuit design ensured the rapid adoption of standardized integrated circuits in place of designs using discrete transistors.

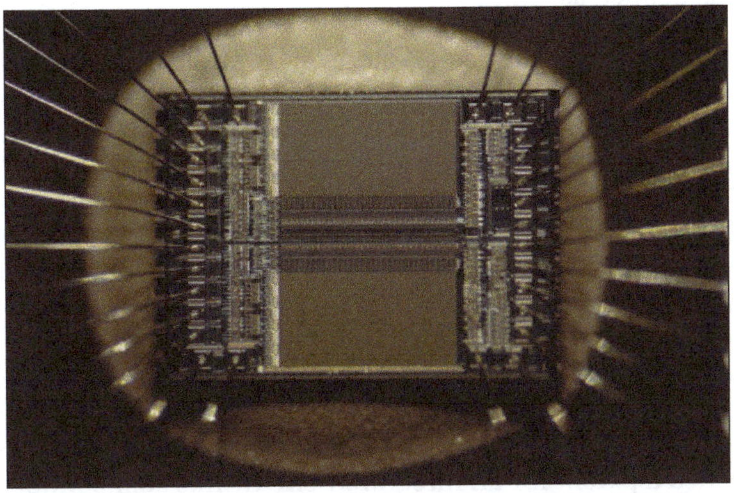

Integrated circuit from an EPROM memory microchip showing the memory blocks, the supporting circuitry and the fine silver wires which connect the integrated circuit die to the legs of the packaging.

ICs have two main advantages over discrete circuits: cost and performance. Cost is low because the chips, with all their components, are printed as a unit by photolithography rather than being constructed one transistor at a time. Furthermore, packaged ICs use much less material than discrete circuits. Performance is high because the IC's components switch quickly and consume little power (compared to their discrete counterparts) as a result of the small size and close proximity of the components. As of 2012, typical chip areas range from a few square millimeters to around 450 mm^2, with up to 9 million transistors per mm^2.

Synthetic detail of an integrated circuit through four layers of planarized copper interconnect, down to the polysilicon (pink), wells (greyish), and substrate (green).

Integrated circuits are used in virtually all electronic equipment today and have revolutionized the world of electronics. Computers, mobile phones, and other digital home appliances are now inextricable parts of the structure of modern societies, made possible by the low cost of ICs.

Terminology

An *integrated circuit* is defined as:

A circuit in which all or some of the circuit elements are inseparably associated and electrically interconnected so that it is considered to be indivisible for the purposes of construction and commerce.

Circuits meeting this definition can be constructed using many different technologies, including thin-film transistor, thick film technology, or hybrid integrated circuit. However, in general usage *integrated circuit* has come to refer to the single-piece circuit construction originally known as a *monolithic integrated circuit*.

Invention

Early developments of the integrated circuit go back to 1949, when German engineer Werner Jacobi (Siemens AG) filed a patent for an integrated-circuit-like semiconductor amplifying device showing five transistors on a common substrate in a 3-stage amplifier arrangement. Jacobi disclosed small and cheap hearing aids as typical industrial applications of his patent. An immediate commercial use of his patent has not been reported.

The idea of the integrated circuit was conceived by Geoffrey W.A. Dummer (1909–2002), a radar scientist working for the Royal Radar Establishment of the British Ministry of Defence. Dummer presented the idea to the public at the Symposium on Progress in Quality Electronic Components in Washington, D.C. on 7 May 1952. He gave many symposia publicly to propagate his ideas, and unsuccessfully attempted to build such a circuit in 1956.

Jack Kilby's original integrated circuit.

A precursor idea to the IC was to create small ceramic squares (wafers), each containing a single miniaturized component. Components could then be integrated and wired into a bidimensional or tridimensional compact grid. This idea, which seemed very promising in 1957, was proposed to the US Army by Jack Kilby and led to the short-lived Micromodule Program (similar to 1951's Project Tinkertoy). However, as the project was gaining momentum, Kilby came up with a new, revolutionary design: the IC.

Newly employed by Texas Instruments, Kilby recorded his initial ideas concerning the integrated circuit in July 1958, successfully demonstrating the first working integrated example on 12 September 1958. In his patent application of 6 February 1959, Kilby described his new device as "a body of semiconductor material ... wherein all the components of the electronic circuit are completely integrated." The first customer for the new invention was the US Air Force.

Kilby won the 2000 Nobel Prize in Physics for his part in the invention of the integrated circuit. His work was named an IEEE Milestone in 2009.

Half a year after Kilby, Robert Noyce at Fairchild Semiconductor developed his own idea of an integrated circuit that solved many practical problems Kilby's had not. Noyce's design was made of silicon, whereas Kilby's chip was made of germanium. Noyce credited Kurt Lehovec of Sprague Electric for the *principle of p–n junction isolation* caused by the action of a biased p–n junction (the diode) as a key concept behind the IC.

Fairchild Semiconductor was also home of the first silicon-gate IC technology with self-aligned gates, the basis of all modern CMOS computer chips. The technology was developed by Italian physicist Federico Faggin in 1968, who later joined Intel in order to develop the very first single-chip Central Processing Unit (CPU) (Intel 4004), for which he received the National Medal of Technology and Innovation in 2010.

Generations

In the early days of simple integrated circuits, the technology's large scale limited each chip to only a few transistors, and the low degree of integration meant the design process was relatively simple. Manufacturing yields were also quite low by today's standards. As the technology progressed, millions, then billions of transistors could be placed on one chip, and good designs required thorough planning, giving rise to new design methods.

Name	Signification	Year	Transistors number	Logic gates number
SSI	*small-scale integration*	1964	1 to 10	1 to 12
MSI	*medium-scale integration*	1968	10 to 500	13 to 99
LSI	*large-scale integration*	1971	500 to 20,000	100 to 9,999
VLSI	*very large-scale integration*	1980	20,000 to 1,000,000	10,000 to 99,999
ULSI	*ultra-large-scale integration*	1984	1,000,000 and more	100,000 and more

SSI, MSI and LSI

The first integrated circuits contained only a few transistors. Early digital circuits containing tens of transistors provided a few logic gates, and early linear ICs such as the Plessey SL201 or the Philips TAA320 had as few as two transistors. The number of transistors in an integrated circuit has increased dramatically since then. The term "large scale integration" (LSI) was first used by IBM scientist Rolf Landauer when describing the theoretical concept; that term gave rise to the terms "small-scale integration" (SSI), "medium-scale integration" (MSI), "very-large-scale integration" (VLSI), and "ultra-large-scale integration" (ULSI). The early integrated circuits were SSI.

SSI circuits were crucial to early aerospace projects, and aerospace projects helped inspire development of the technology. Both the Minuteman missile and Apollo program needed lightweight digital computers for their inertial guidance systems. Although the Apollo guidance computer led and motivated integrated-circuit technology, it was the Minuteman missile that forced it into mass-production. The Minuteman missile program and various other Navy programs accounted for the total $4 million integrated circuit market in 1962, and by 1968, U.S. Government space and defense spending still accounted for 37% of the $312 million total production. The demand by the U.S. Government supported the nascent integrated circuit market until costs fell enough to allow firms to penetrate the industrial, and eventually, the consumer markets. The average price per integrated circuit dropped from $50.00 in 1962 to $2.33 in 1968. Integrated circuits began to appear in consumer products by the turn of the decade, a typical application being FM inter-carrier sound processing in television receivers.

The first MOS chips were small-scale integration chips for NASA satellites.

The next step in the development of integrated circuits, taken in the late 1960s, introduced devices which contained hundreds of transistors on each chip, called "medium-scale integration" (MSI).

In 1964, Frank Wanlass demonstrated a single-chip 16-bit shift register he designed, with an incredible (at the time) 120 transistors on a single chip.

MSI devices were attractive economically because while they cost little more to produce than SSI devices, they allowed more complex systems to be produced using smaller circuit boards, less assembly work (because of fewer separate components), and a number of other advantages.

Further development, driven by the same economic factors, led to "large-scale integration" (LSI) in the mid-1970s, with tens of thousands of transistors per chip.

SSI and MSI devices often were manufactured by masks created by hand-cutting Rubylith. An engineer would inspect and verify the completeness of each mask. LSI devices contain so many transistors, interconnecting wires, and other features that it is considered impossible for a human to check the masks or even do the original design entirely by hand. The engineer depends on computer programs and other hardware aids to do most of this work.

Integrated circuits such as 1K-bit RAMs, calculator chips, and the first microprocessors, that began to be manufactured in moderate quantities in the early 1970s, had under 4000 transistors. True LSI circuits, approaching 10,000 transistors, began to be produced around 1974, for computer main memories and second-generation microprocessors.

VLSI

The final step in the development process, starting in the 1980s and continuing through the present, was "very-large-scale integration" (VLSI). The development started with hundreds of thousands of transistors in the early 1980s, and continues beyond several billion transistors as of 2009.

Multiple developments were required to achieve this increased density. Manufacturers moved to smaller design rules and cleaner fabrication facilities, so that they could make chips with more transistors and maintain adequate yield. The path of process improvements was summarized by the International Technology Roadmap for Semiconductors (ITRS). Design tools improved enough to make it practical to finish these designs in a reasonable time. The more energy-efficient CMOS replaced NMOS and PMOS, avoiding a prohibitive increase in power consumption.

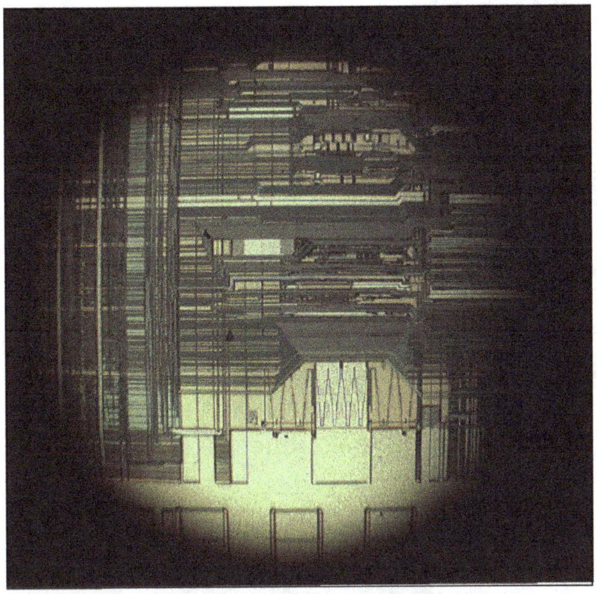

Upper interconnect layers on an Intel 80486DX2 microprocessor die.

In 1986 the first one-megabit RAM chips were introduced, containing more than one million transistors. Microprocessor chips passed the million-transistor mark in 1989 and the billion-transistor mark in 2005. The trend continues largely unabated, with chips introduced in 2007 containing tens of billions of memory transistors.

ULSI, WSI, SOC and 3D-IC

To reflect further growth of the complexity, the term *ULSI* that stands for "ultra-large-scale integration" was proposed for chips of more than 1 million transistors.

Wafer-scale integration (WSI) is a means of building very large integrated circuits that uses an entire silicon wafer to produce a single "super-chip". Through a combination of large size and reduced packaging, WSI could lead to dramatically reduced costs for some systems, notably massively parallel supercomputers. The name is taken from the term Very-Large-Scale Integration, the current state of the art when WSI was being developed.

A system-on-a-chip (SoC or SOC) is an integrated circuit in which all the components needed for a computer or other system are included on a single chip. The design of such a device can be complex and costly, and building disparate components on a single piece of silicon may compromise the efficiency of some elements. However, these drawbacks are offset by lower manufacturing and assembly costs and by a greatly reduced power budget: because signals among the components are kept on-die, much less power is required.

A three-dimensional integrated circuit (3D-IC) has two or more layers of active electronic components that are integrated both vertically and horizontally into a single circuit. Communication between layers uses on-die signaling, so power consumption is much lower than in equivalent separate circuits. Judicious use of short vertical wires can substantially reduce overall wire length for faster operation.

Advances in Integrated Circuits

ICs have consistently migrated to smaller feature sizes over the years, allowing more circuitry to be packed on each chip. This increased capacity per unit area can be used to decrease cost or increase functionality—see Moore's law which, in its modern interpretation, states that the number of transistors in an integrated circuit doubles every two years. In general, as the feature size shrinks, almost everything improves—the cost per transistor and the switching power consumption per transistor go down, and the speed goes up. However, ICs with nanometer-scale devices are not without their problems, principal among which is leakage current, although innovations in high-κ dielectrics aim to solve these problems. Since these speed and power consumption gains are apparent to the end user, there is fierce competition among the manufacturers to use finer geometries. This process, and the expected progress over the next few years, is described by the International Technology Roadmap for Semiconductors (ITRS).

The die from an Intel 8742, an 8-bit microcontroller that includes a CPU running at 12 MHz, 128 bytes of RAM, 2048 bytes of EPROM, and I/O in the same chip.

Among the most advanced integrated circuits are the microprocessors or "cores", which control everything from computers and cellular phones to digital microwave ovens. Digital memory chips and application-specific integrated circuits (ASICs) are examples of other families of integrated circuits that are important to the modern information society. While the cost of designing and developing a complex integrated circuit is quite high, when spread across typically millions of production units the individual IC cost is minimized. The performance of ICs is high because the small size allows short traces which in turn allows low power logic (such as CMOS) to be used at fast switching speeds.

In current research projects, integrated circuits are also developed for sensoric applications in medical implants or other bioelectronic devices. Special sealing techniques have to be applied in such biogenic environments to avoid corrosion or biodegradation of the exposed semiconductor materials. As one of the few materials well established in CMOS technology, titanium nitride (TiN) turned out as exceptionally stable and well suited for electrode applications in medical implants.

Classification

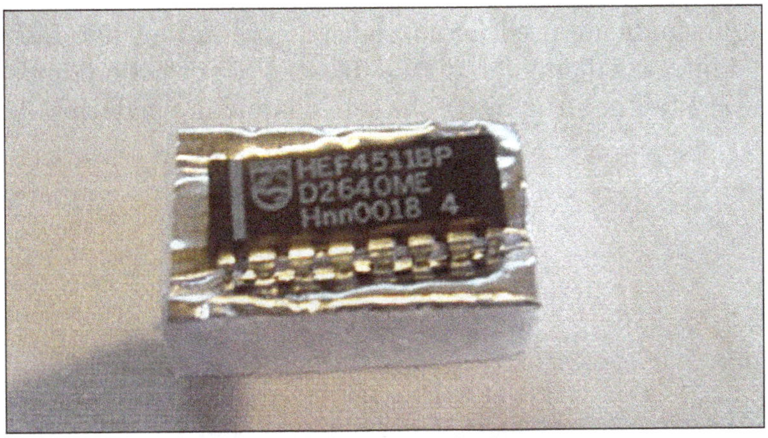

A CMOS 4511 IC in a DIP.

Integrated circuits can be classified into analog, digital and mixed signal (both analog and digital on the same chip).

Digital integrated circuits can contain anywhere from one to millions of logic gates, flip-flops, multiplexers, and other circuits in a few square millimeters. The small size of these circuits allows high speed, low power dissipation, and reduced manufacturing cost compared with board-level integration. These digital ICs, typically microprocessors, DSPs, and microcontrollers, work using binary mathematics to process "one" and "zero" signals.

Analog ICs, such as sensors, power management circuits, and operational amplifiers, work by processing continuous signals. They perform functions like amplification, active filtering, demodulation, and mixing. Analog ICs ease the burden on circuit designers by having expertly designed analog circuits available instead of designing a difficult analog circuit from scratch.

ICs can also combine analog and digital circuits on a single chip to create functions such as A/D converters and D/A converters. Such mixed-signal circuits offer smaller size and lower cost, but must carefully account for signal interference.

Modern electronic component distributors often further sub-categorize the huge variety of integrated circuits now available:

- Digital ICs are further sub-categorized as logic ICs, memory chips, interface ICs (level shifters, serializer/deserializer, etc.), Power Management ICs, and programmable devices.
- Analog ICs are further sub-categorized as linear ICs and RF ICs.
- mixed-signal integrated circuits are further sub-categorized as data acquisition ICs (including A/D converters, D/A converter, digital potentiometers) and clock/timing ICs.

Manufacturing

Fabrication

The semiconductors of the periodic table of the chemical elements were identified as the most likely materials for a *solid-state vacuum tube*. Starting with copper oxide, proceeding to germanium, then silicon, the materials were systematically studied in the 1940s and 1950s. Today, monocrystalline silicon is the main substrate used for ICs although some III-V compounds of the periodic table such as gallium arsenide are used for specialized applications like LEDs, lasers, solar cells and the highest-speed integrated circuits. It took decades to perfect methods of creating crystals without defects in the crystalline structure of the semiconducting material.

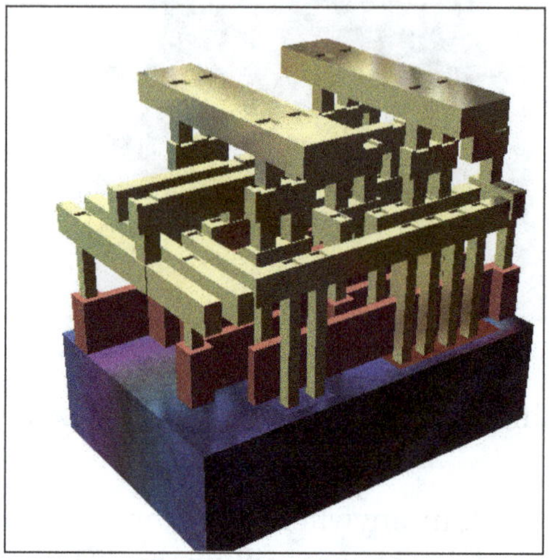

Rendering of a small standard cell with three metal layers (dielectric has been removed). The sand-colored structures are metal interconnect, with the vertical pillars being contacts, typically plugs of tungsten. The reddish structures are polysilicon gates, and the solid at the bottom is the crystalline silicon bulk.

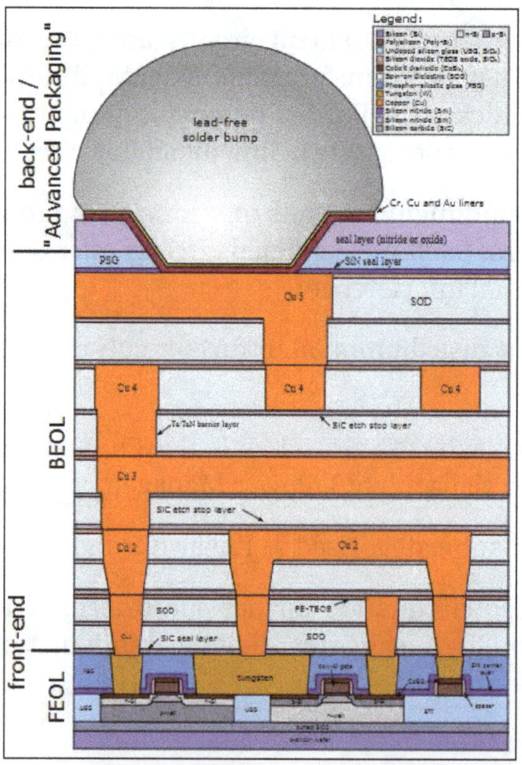

Schematic structure of a CMOS chip, as built in the early 2000s. The graphic shows LDD-MISFET's on an SOI substrate with five metallization layers and solder bump for flip-chip bonding. It also shows the section for FEOL (front-end of line), BEOL (back-end of line) and first parts of back-end process.

Semiconductor ICs are fabricated in a planar process which includes three key process steps – imaging, deposition and etching. The main process steps are supplemented by doping and cleaning.

Mono-crystal silicon wafers (or for special applications, silicon on sapphire or gallium arsenide wafers) are used as the *substrate*. Photolithography is used to mark different areas of the substrate to be doped or to have polysilicon, insulators or metal (typically aluminium) tracks deposited on them.

- Integrated circuits are composed of many overlapping layers, each defined by photolithography, and normally shown in different colors. Some layers mark where various dopants are diffused into the substrate (called diffusion layers), some define where additional ions are implanted (implant layers), some define the conductors (polysilicon or metal layers), and some define the connections between the conducting layers (via or contact layers). All components are constructed from a specific combination of these layers.
- In a self-aligned CMOS process, a transistor is formed wherever the gate layer (polysilicon or metal) crosses a diffusion layer.
- Capacitive structures, in form very much like the parallel conducting plates of a traditional electrical capacitor, are formed according to the area of the "plates", with insulating material between the plates. Capacitors of a wide range of sizes are common on ICs.
- Meandering stripes of varying lengths are sometimes used to form on-chip resistors, though most logic circuits do not need any resistors. The ratio of the length of the resistive structure to its width, combined with its sheet resistivity, determines the resistance.
- More rarely, inductive structures can be built as tiny on-chip coils, or simulated by gyrators.

Since a CMOS device only draws current on the *transition* between logic states, CMOS devices consume much less current than bipolar devices.

A random access memory is the most regular type of integrated circuit; the highest density devices are thus memories; but even a microprocessor will have memory on the chip. Although the structures are intricate – with widths which have been shrinking for decades – the layers remain much thinner than the device widths. The layers of material are fabricated much like a photographic process, although light waves in the visible spectrum cannot be used to "expose" a layer of material, as they would be too large for the features. Thus photons of higher frequencies (typically ultraviolet) are used to create the patterns for each layer. Because each feature is so small, electron microscopes are essential tools for a pro-cess engineer who might be debugging a fabrication process.

Each device is tested before packaging using automated test equipment (ATE), in a process known as wafer testing, or wafer probing. The wafer is then cut into rectangular blocks, each of which is called a *die*. Each good die (plural *dice*, *dies*, or *die*) is then connected into a package using aluminium (or gold) bond wires which are thermosonically bonded to *pads*, usually found around the edge of the die. . Thermosonic bonding was first introduced by A. Coucoulas which provided a reliable means of forming these vital electrical connections to the outside world. After packaging, the devices go through final testing on the same or similar ATE used during wafer probing. Industrial CT scanning can also be used. Test cost can account for over 25% of the cost of fabrication on lower-cost products, but can be negligible on low-yielding, larger, or higher-cost devices.

As of 2005, a fabrication facility (commonly known as a *semiconductor fab*) costs over US$1 billion to construct. The cost of a fabrication facility rises over time (Rock's law) because much of the operation is automated. Today, the most advanced processes employ the following techniques:

- The wafers are up to 300 mm in diameter (wider than a common dinner plate).
- Use of 32 nanometer or smaller chip manufacturing process. Intel, IBM, NEC, and AMD are using ~32 nanometers for their CPU chips. IBM and AMD introduced immersion lithography for their 45 nm processes.
- Copper interconnects where copper wiring replaces aluminium for interconnects.
- Low-K dielectric insulators.
- Silicon on insulator (SOI).
- Strained silicon in a process used by IBM known as strained silicon directly on insulator (SSDOI).
- Multigate devices such as tri-gate transistors being manufactured by Intel from 2011 in their 22 nm process.

Packaging

A Soviet MSI nMOS chip made in 1977, part of a four-chip calculator set designed in 1970.

The earliest integrated circuits were packaged in ceramic flat packs, which continued to be used by the military for their reliability and small size for many years. Commercial circuit packaging quickly moved to the dual in-line package (DIP), first in ceramic and later in plastic. In the 1980s pin counts of VLSI circuits exceeded the practical limit for DIP packaging, leading to pin grid array (PGA) and leadless chip carrier (LCC) packages. Surface mount packaging appeared in the early

1980s and became popular in the late 1980s, using finer lead pitch with leads formed as either gull-wing or J-lead, as exemplified by small-outline integrated circuit – a carrier which occupies an area about 30–50% less than an equivalent DIP, with a typical thickness that is 70% less. This package has "gull wing" leads protruding from the two long sides and a lead spacing of 0.050 inches.

In the late 1990s, plastic quad flat pack (PQFP) and thin small-outline package (TSOP) packages became the most common for high pin count devices, though PGA packages are still often used for high-end microprocessors. Intel and AMD are currently transitioning from PGA packages on high-end microprocessors to land grid array (LGA) packages.

Ball grid array (BGA) packages have existed since the 1970s. Flip-chip Ball Grid Array packages, which allow for much higher pin count than other package types, were developed in the 1990s. In an FCBGA package the die is mounted upside-down (flipped) and connects to the package balls via a package substrate that is similar to a printed-circuit board rather than by wires. FCBGA packages allow an array of input-output signals (called Area-I/O) to be distributed over the entire die rather than being confined to the die periphery.

Traces out of the die, through the package, and into the printed circuit board have very different electrical properties, compared to on-chip signals. They require special design techniques and need much more electric power than signals confined to the chip itself.

When multiple dies are put in one package, it is called SiP, for *System In Package*. When multiple dies are combined on a small substrate, often ceramic, it's called an MCM, or Multi-Chip Module. The distinction between a big MCM and a small printed circuit board is sometimes fuzzy.

Chip Labeling and Manufacture Date

Most integrated circuits large enough to include identifying information include four common sections: the manufacturer's name or logo, the part number, a part production batch number and serial number, and a four-digit code that identifies when the chip was manufactured. Extremely small surface mount technology parts often bear only a number used in a manufacturer's lookup table to find the chip characteristics.

The manufacturing date is commonly represented as a two-digit year followed by a two-digit week code, such that a part bearing the code 8341 was manufactured in week 41 of 1983, or approximately in October 1983.

Intellectual Property

The possibility of copying by photographing each layer of an integrated circuit and preparing photomasks for its production on the basis of the photographs obtained is the main reason for the introduction of legislation for the protection of layout-designs.The Semiconductor Chip Protection Act (SCPA) of 1984 established a new type of intellectual property protection for mask works that are fixed in semiconductor chips. It did so by amending title 17 of the United States chapter 9.

A diplomatic conference was held at Washington, D.C., in 1989, which adopted a Treaty on Intellectual Property in Respect of Integrated Circuits (IPIC Treaty).

The Treaty on Intellectual Property in respect of Integrated Circuits, also called Washington Treaty or IPIC Treaty (signed at Washington on 26 May 1989) is currently not in force, but was partially integrated into the TRIPS agreement.

National laws protecting IC layout designs have been adopted in a number of countries.

Other Developments

In the 1980s, programmable logic devices were developed. These devices contain circuits whose logical function and connectivity can be programmed by the user, rather than being fixed by the integrated circuit manufacturer. This allows a single chip to be programmed to implement different LSI-type functions such as logic gates, adders and registers. Current devices called field-programmable gate arrays can now implement tens of thousands of LSI circuits in parallel and operate up to 1.5 GHz.

The techniques perfected by the integrated circuits industry over the last three decades have been used to create very small mechanical devices driven by electricity using a technology known as microelectromechanical systems. These devices are used in a variety of commercial and military applications. Example commercial applications include DLP projectors, inkjet printers, and accelerometers and MEMS gyroscopes used to deploy automobile airbags.

As of 2014, the vast majority of all transistors are fabricated in a single layer on one side of a chip of silicon in a flat 2-dimensional planar process. Researchers have produced prototypes of several promising alternatives, such as:

- Fabricating transistors over the entire surface of a small sphere of silicon.
- Various approaches to stacking several layers of transistors to make a three-dimensional integrated circuit, such as through-silicon via, "monolithic 3D", stacked wire bonding, etc.
- Transistors built from other materials: graphene transistors, molybdenite transistors, carbon nanotube field-effect transistor, gallium nitride transistor, transistor-like nanowire electronic devices, organic field-effect transistor, etc.
- Modifications to the substrate, typically to make "flexible transistors" for a flexible display or other flexible electronics, possibly leading to a roll-away computer.

In the past, radios could not be fabricated in the same low-cost processes as microprocessors. But since 1998, a large number of radio chips have been developed using CMOS processes. Examples include Intel's DECT cordless phone, or Atheros's 802.11 card.

Future developments seem to follow the multi-core multi-microprocessor paradigm, already used by the Intel and AMD dual-core processors. Rapport Inc. and IBM started shipping the KC256 in 2006, a 256-core microprocessor. Intel, as recently as February–August 2011, unveiled a prototype, "not for commercial sale" chip that bears 80 cores. Each core is capable of handling its own task independently of the others. This is in response to the heat-versus-speed limit that is about to be reached using existing transistor technology. This design provides a new challenge to chip programming. Parallel programming languages such as the open-source X10 programming language are designed to assist with this task.

Since the early 2000s, the integration of optical functionality (optical computing) into silicon chips has been actively pursued in both academic research and in industry resulting in the successful commercialization of silicon based integrated optical transceivers combining optical devices (modulators, detectors, routing) with CMOS based electronics.

Silicon Labelling and Graffiti

To allow identification during production most silicon chips will have a serial number in one corner. It is also common to add the manufacturer's logo. Ever since ICs were created, some chip designers have used the silicon surface area for surreptitious, non-functional images or words. These are sometimes referred to as chip art, silicon art, silicon graffiti or silicon doodling.

Processes of Integrated Circuit

Very-large-scale Integration

Very-large-scale integration (VLSI) is the process of creating an integrated circuit (IC) by combining thousands of transistors into a single chip. VLSI began in the 1970s when complex semiconductor and communication technologies were being developed. The microprocessor is a VLSI device. Before the introduction of VLSI technology most ICs had a limited set of functions they could perform. An electronic circuit might consist of a CPU, ROM, RAM and other glue logic. VLSI lets IC designers add all of these into one chip.

A VLSI integrated-circuit die.

History

During the mid-1920s, several inventors attempted devices that were intended to control current in solid-state diodes and convert them into triodes. Success did not come until after WWII, during which the attempt to improve silicon and germanium crystals for use as radar detectors led to improvements in fabrication and in the understanding of quantum mechanical states of carriers in semiconductors. Then scientists who had been diverted to radar development returned to solid-state device development. With the invention of transistors at Bell Labs in 1947, the field of electronics shifted from vacuum tubes to solid-state devices.

With the small transistor at their hands, electrical engineers of the 1950s saw the possibilities of constructing far more advanced circuits. As the complexity of circuits grew, problems arose.

One problem was the size of the circuit. A complex circuit, like a computer, was dependent on speed. If the components of the computer were too large or the wires interconnecting them too long, the electric signals couldn't travel fast enough through the circuit, thus making the computer too slow to be effective.

Jack Kilby at Texas Instruments found a solution to this problem in 1958. Kilby's idea was to make all the components and the chip out of the same block (monolith) of semiconductor material. Kilby presented his idea to his superiors, and was allowed to build a test version of his circuit. In September 1958, he had his first integrated circuit ready. Although the first integrated circuit was crude and had some problems, the idea was groundbreaking. By making all the parts out of the same block of material and adding the metal needed to connect them as a layer on top of it, there was no need for discrete components. No more wires and components had to be assembled manually. The circuits could be made smaller, and the manufacturing process could be automated. From here, the idea of integrating all components on a single silicon wafer came into existence, which led to development in small-scale integration (SSI) in the early 1960s, medium-scale integration (MSI) in the late 1960s, and then large-scale integration (LSI) as well as VLSI in the 1970s and 1980s, with tens of thousands of transistors on a single chip (later hundreds of thousands, then millions, and now billions (10^9)).

Developments

The first semiconductor chips held two transistors each. Subsequent advances added more transistors, and as a consequence, more individual functions or systems were integrated over time. The first integrated circuits held only a few devices, perhaps as many as ten diodes, transistors, resistors and capacitors, making it possible to fabricate one or more logic gates on a single device. Now known retrospectively as *small-scale integration* (SSI), improvements in technique led to devices with hundreds of logic gates, known as *medium-scale integration* (MSI). Further improvements led to *large-scale integration* (LSI), i.e. systems with at least a thousand logic gates. Current technology has moved far past this mark and today's microprocessors have many millions of gates and billions of individual transistors.

At one time, there was an effort to name and calibrate various levels of large-scale integration above VLSI. Terms like *ultra-large-scale integration* (ULSI) were used. But the huge number of gates and transistors available on common devices has rendered such fine distinctions moot. Terms suggesting greater than VLSI levels of integration are no longer in widespread use.

As of early 2008, billion-transistor processors are commercially available. This became more commonplace as semiconductor fabrication advanced from the then-current generation of 65 nm processes. Current designs, unlike the earliest devices, use extensive design automation and automated logic synthesis to lay out the transistors, enabling higher levels of complexity in the resulting logic functionality. Certain high-performance logic blocks like the SRAM (static random-access memory) cell, are still designed by hand to ensure the highest efficiency.

Structured Design

Structured VLSI design is a modular methodology originated by Carver Mead and Lynn Conway for saving microchip area by minimizing the interconnect fabrics area. This is obtained by repetitive arrangement of rectangular macro blocks which can be interconnected using wiring by abutment. An example is partitioning the layout of an adder into a row of equal bit slices cells. In complex designs this structuring may be achieved by hierarchical nesting.

Structured VLSI design had been popular in the early 1980s, but lost its popularity later because of the advent of placement and routing tools wasting a lot of area by routing, which is tolerated because of the progress of Moore's Law. When introducing the hardware description language KARL in the mid' 1970s, Reiner Hartenstein coined the term "structured VLSI design" (originally as "structured LSI design"), echoing Edsger Dijkstra's structured programming approach by procedure nesting to avoid chaotic *spaghetti-structured* program.

Challenges

As microprocessors become more complex due to technology scaling, microprocessor designers have encountered several challenges which force them to think beyond the design plane, and look ahead to post-silicon:

- Process variation – As photolithography techniques get closer to the fundamental laws of optics, achieving high accuracy in doping concentrations and etched wires is becoming more difficult and prone to errors due to variation. Designers now must simulate across multiple fabrication process corners before a chip is certified ready for production, or use system-level techniques for dealing with effects of variation.

- Stricter design rules – Due to lithography and etch issues with scaling, design rules for layout have become increasingly stringent. Designers must keep ever more of these rules in mind while laying out custom circuits. The overhead for custom design is now reaching a tipping point, with many design houses opting to switch to electronic design automation (EDA) tools to automate their design process.

- Timing/design closure – As clock frequencies tend to scale up, designers are finding it more difficult to distribute and maintain low clock skew between these high frequency clocks across the entire chip. This has led to a rising interest in multicore and multiprocessor architectures, since an overall speedup can be obtained even with lower clock frequency by using the computational power of all the cores.

- First-pass success – As die sizes shrink (due to scaling), and wafer sizes go up (due to lower manufacturing costs), the number of dies per wafer increases, and the complexity of making suitable photomasks goes up rapidly. A mask set for a modern technology can cost several million dollars. This non-recurring expense deters the old iterative philosophy involving several "spin-cycles" to find errors in silicon, and encourages first-pass silicon success. Several design philosophies have been developed to aid this new design flow, including design for manufacturing (DFM), design for test (DFT), and Design for X.

Semiconductor Device Fabrication

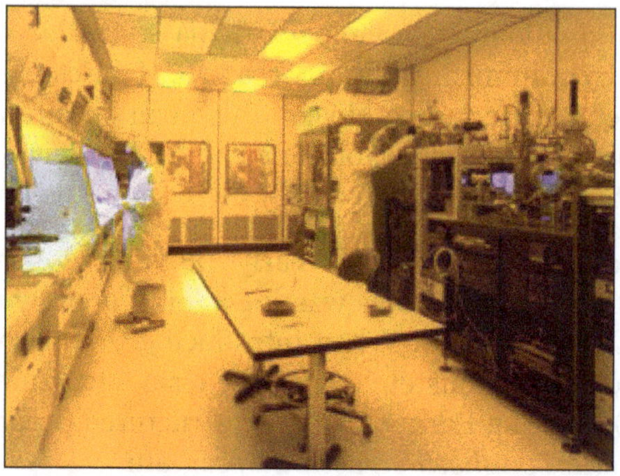

NASA's Glenn Research Center clean room.

Semiconductor device fabrication is the process used to create the integrated circuits that are present in everyday electrical and electronic devices. It is a multiple-step sequence of photo lithographic and chemical processing steps during which electronic circuits are gradually created on a wafer made of pure semiconducting material. Silicon is almost always used, but various compound semiconductors are used for specialized applications.

The entire manufacturing process, from start to packaged chips ready for shipment, takes six to eight weeks and is performed in highly specialized facilities referred to as fabs.

History

When feature widths were far greater than about 10 micrometres, purity was not the issue that it is today in device manufacturing. As devices became more integrated, cleanrooms became even cleaner. Today, the fabs are pressurized with filtered air to remove even the smallest particles, which could come to rest on the wafers and contribute to defects. The workers in a semiconductor fabrication facility are required to wear cleanroom suits to protect the devices from human contamination.

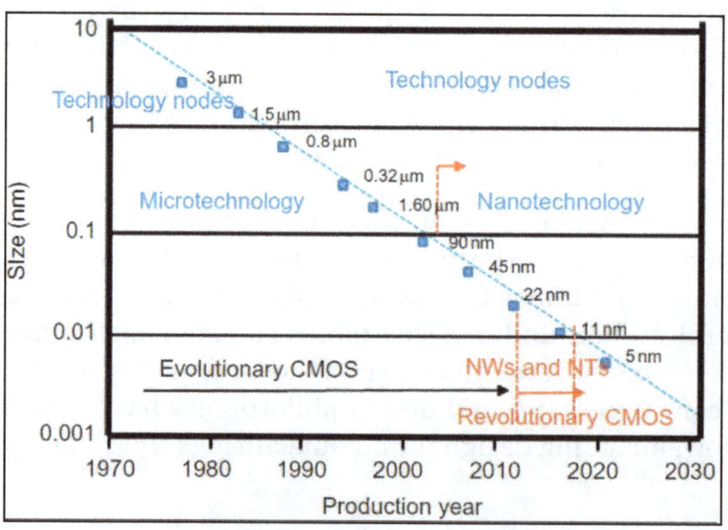

Semiconductor device manufacturing has spread from Texas and California in the 1960s to the rest of the world, including Europe, the Middle East, and Asia. It is a global business today. The leading semiconductor manufacturers typically have facilities all over the world. Intel, the world's largest manufacturer, has facilities in Europe and Asia as well as the U.S. Other top manufacturers include Taiwan Semiconductor Manufacturing Company (Taiwan), United Microelectronics Corporation (Taiwan), STMicroelectronics (Europe), Analog Devices (US), Integrated Device Technology (US), Atmel (US/Europe), Freescale Semiconductor (US), Samsung (Korea), Texas Instruments (US), IBM (US), GlobalFoundries (Germany, Singapore, US), Toshiba (Japan), NEC Electronics (Japan), Infineon (Europe, US, Asia), Renesas (Japan), Fujitsu (Japan/US), NXP Semiconductors (Europe, Asia and US), Micron Technology (US), Hynix (Korea), and SMIC (China).

Wafers

A typical wafer is made out of extremely pure silicon that is grown into mono-crystalline cylindrical ingots (boules) up to 300 mm (slightly less than 12 inches) in diameter using the Czochralski process. These ingots are then sliced into wafers about 0.75 mm thick and polished to obtain a very regular and flat surface.

Processing

In semiconductor device fabrication, the various processing steps fall into four general categories: deposition, removal, patterning, and modification of electrical properties.

- *Deposition* is any process that grows, coats, or otherwise transfers a material onto the wafer. Available technologies include physical vapor deposition (PVD), chemical vapor deposition (CVD), electrochemical deposition (ECD), molecular beam epitaxy (MBE) and more recently, atomic layer deposition (ALD) among others.
- *Removal* is any process that removes material from the wafer; examples include etch processes (either wet or dry) and chemical-mechanical planarization (CMP).
- *Patterning* is the shaping or altering of deposited materials, and is generally referred to as lithography. For example, in conventional lithography, the wafer is coated with a chemical called a *photoresist*; then, a machine called a *stepper* focuses, aligns, and moves a mask, exposing select portions of the wafer below to short wavelength light; the exposed regions are washed away by a developer solution. After etching or other processing, the remaining photoresist is removed by plasma ashing.
- *Modification of electrical properties* has historically entailed doping transistor *sources* and *drains* (originally by diffusion furnaces and later by ion implantation). These doping processes are followed by furnace annealing or, in advanced devices, by rapid thermal annealing (RTA); annealing serves to activate the implanted dopants. Modification of electrical properties now also extends to the reduction of a material's dielectric constant in low-k insulators via exposure to ultraviolet light in UV processing (UVP).

Front-end-of-line (FEOL) Processing

FEOL processing refers to the formation of the transistors directly in the silicon. The raw wafer is engineered by the growth of an ultrapure, virtually defect-free silicon layer through epitaxy. In the most advanced logic devices, *prior* to the silicon epitaxy step, tricks are performed to improve the performance of the transistors to be built. One method involves introducing a *straining step* wherein a silicon variant such as silicon-germanium (SiGe) is deposited. Once the epitaxial silicon is deposited, the crystal lattice becomes stretched somewhat, resulting in improved electronic mobility. Another method, called *silicon on insulator* technology involves the insertion of an insulating layer between the raw silicon wafer and the thin layer of subsequent silicon epitaxy. This method results in the creation of transistors with reduced parasitic effects.

Gate Oxide and Implants

Front-end surface engineering is followed by growth of the gate dielectric (traditionally silicon dioxide), patterning of the gate, patterning of the source and drain regions, and subsequent implantation or diffusion of dopants to obtain the desired complementary electrical properties. In dynamic random-access memory (DRAM) devices, storage capacitors are also fabricated at this time, typically stacked above the access transistor (the now defunct DRAM manufacturer Qimonda implemented these capacitors with trenches etched deep into the silicon surface).

Metal Layers

Once the various semiconductor devices have been created, they must be interconnected to form the desired electrical circuits. This occurs in a series of wafer processing steps collectively referred to as BEOL. BEOL processing involves creating metal interconnecting wires that are isolated by dielectric layers. The insulating material has traditionally been a form of SiO_2 or a silicate glass, but recently new low dielectric constant materials are being used (such as silicon oxycarbide), typically providing dielectric constants around 2.7 (compared to 3.9 for SiO_2), although materials with con-stants as low as 2.2 are being offered to chipmakers.

Interconnect

Synthetic detail of a standard cell through four layers of planarized copper interconnect, down to the polysilicon (pink), wells (greyish) and substrate (green).

Historically, the metal wires have been composed of aluminum. In this approach to wiring (often called *subtractive aluminum*), blanket films of aluminum are deposited first, patterned, and then etched, leaving isolated wires. Dielectric material is then deposited over the exposed wires. The various metal layers are interconnected by etching holes (called *"vias")* in the insulating material and then depositing tungsten in them with a CVD technique; this approach is still used in the fabrication of many memory chips such as dynamic random-access memory (DRAM), because the number of interconnect levels is small (currently no more than four).

More recently, as the number of interconnect levels for logic has substantially increased due to the large number of transistors that are now interconnected in a modern microprocessor, the timing delay in the wiring has become so significant as to prompt a change in wiring material (from aluminum to copper layer) and a change in dielectric material (from silicon dioxides to newer low-K insulators). This performance enhancement also comes at a reduced cost via damascene processing, which eliminates processing steps. As the number of interconnect levels increases, planarization of the previous layers is required to ensure a flat surface prior to subsequent lithography. Without it, the levels would become increasingly crooked, extending outside the depth of focus of available lithography, and thus interfering with the ability to pattern. CMP (chemical-mechanical planarization) is the primary processing method to achieve such planarization, although dry *etch back* is still sometimes employed when the number of interconnect levels is no more than three.

Wafer Test

The highly serialized nature of wafer processing has increased the demand for metrology in between the various processing steps. For example, thin film metrology based on ellipsometry or reflectometry is used to tightly control the thickness of gate oxide, as well as the thickness, refractive index and extinction coefficient of photoresist and other coatings. Wafer test metrology equipment is used to verify that the wafers haven't been damaged by previous processing steps up until testing; if too many dies on one wafer have failed, the entire wafer is scrapped to avoid the costs of further processing. Virtual metrology has been used to predict wafer properties based on statistical methods without performing the physical measurement itself.

Device Test

Once the front-end process has been completed, the semiconductor devices are subjected to a variety of electrical tests to determine if they function properly. The proportion of devices on the wafer found to perform properly is referred to as the yield. Manufacturers are typically secretive about their yields, but it can be as low as 30%. Process variation is one among many reasons for low yield.

The fab tests the chips on the wafer with an electronic tester that presses tiny probes against the chip. The machine marks each bad chip with a drop of dye. Currently, electronic dye marking is possible if wafer test data is logged into a central computer database and chips are "binned" (i.e. sorted into virtual bins) according to the predetermined test limits. The resulting binning data can be graphed, or logged, on a wafer map to trace manufacturing defects and mark bad chips. This map can also be used during wafer assembly and packaging.

Chips are also tested again after packaging, as the bond wires may be missing, or analog performance may be altered by the package. This is referred to as the "final test".

Usually, the fab charges for testing time, with prices in the order of cents per second. Testing times vary from a few milliseconds to a couple of seconds, and the test software is optimized for reduced testing time. Multiple chip (multi-site) testing is also possible, because many testers have the resources to perform most or all of the tests in parallel.

Chips are often designed with "testability features" such as scan chains or a "built-in self-test" to speed testing, and reduce testing costs. In certain designs that use specialized analog fab processes, wafers are also laser-trimmed during the testing, in order to achieve tightly-distributed resistance values as specified by the design.

Good designs try to test and statistically manage *corners* (extremes of silicon behavior caused by a high operating temperature combined with the extremes of fab processing steps). Most designs cope with at least 64 corners.

Die Preparation

Once tested, a wafer is typically reduced in thickness before the wafer is scored and then broken into individual dice, a process known as wafer dicing. Only the good, unmarked chips are packaged.

Packaging

Plastic or ceramic packaging involves mounting the die, connecting the die pads to the pins on the package, and sealing the die. Tiny wires are used to connect the pads to the pins. In the old days, wires were attached by hand, but now specialized machines perform the task. Traditionally, these wires have been composed of gold, leading to a lead frame (pronounced "leed frame") of solder-plated copper; lead is poisonous, so lead-free "lead frames" are now mandated by RoHS.

Chip scale package (CSP) is another packaging technology. A plastic dual in-line package, like most packages, is many times larger than the actual die hidden inside, whereas CSP chips are nearly the size of the die; a CSP can be constructed for each die *before* the wafer is diced.

The packaged chips are retested to ensure that they were not damaged during packaging and that the die-to-pin interconnect operation was performed correctly. A laser then etches the chip's name and numbers on the package.

List of Steps

This is a list of processing techniques that are employed numerous times throughout the construction of a modern electronic device; this list does not necessarily imply a specific order.

- Wafer Processing:
  - Wet cleans:
    - Cleaning by solvents such as acetone, trichloroethylene
    - Piranha solution
    - RCA clean

- Photolithography
- Ion implantation (in which dopants are embedded in the wafer creating regions of increased (or decreased) conductivity)
- Dry etching
- Wet etching
- Plasma ashing
- Thermal treatments:
    - Rapid thermal anneal
    - Furnace anneals
    - Thermal oxidation
- Chemical vapor deposition (CVD)
- Physical vapor deposition (PVD)
- Molecular beam epitaxy (MBE)
- Electrochemical deposition (ECD)
- Chemical-mechanical planarization (CMP)
- Wafer testing (where the electrical performance is verified)
- Wafer backgrinding (to reduce the thickness of the wafer so the resulting chip can be put into a thin device like a smartcard or PCMCIA card.)

- Die Preparation:
    - Wafer mounting
    - Die cutting
- IC Packaging:
    - Die attachment
    - IC bonding:
        - Wire bonding
        - Thermosonic bonding
        - Flip chip
        - Wafer bonding

- Tape Automated Bonding (TAB)
- IC encapsulation:
    - Baking
    - Plating
    - Lasermarking
    - Trim and form
- IC Testing

Hazardous Materials

Many toxic materials are used in the fabrication process. These include:

- Poisonous elemental dopants, such as arsenic, antimony, and phosphorus.
- Poisonous compounds, such as arsine, phosphine, and silane.
- Highly reactive liquids, such as hydrogen peroxide, fuming nitric acid, sulfuric acid, and hydrofluoric acid.

It is vital that workers should not be directly exposed to these dangerous substances. The high degree of automation common in the IC fabrication industry helps to reduce the risks of exposure. Most fabrication facilities employ exhaust management systems, such as wet scrubbers, combustors, heated absorber cartridges, etc., to control the risk to workers and to the environment.

Integrated Circuit Packaging

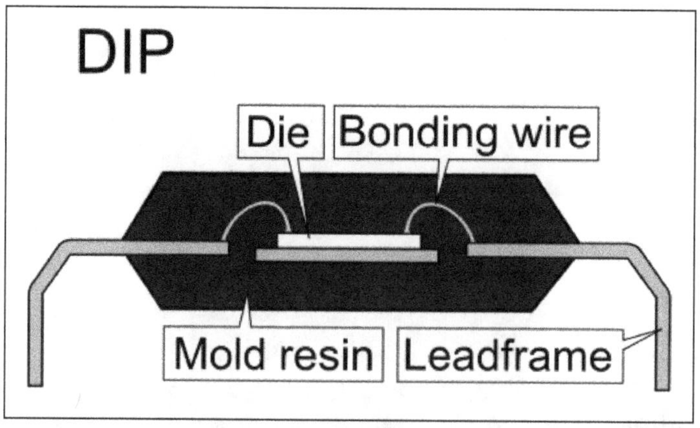

Cross section of a dual in-line package. This type of package houses a small semiconducting die, with nanowires attaching the die to the lead frames, allowing for electrical connections to be made to a PCB.

In electronics manufacturing, integrated circuit packaging is the final stage of semiconductor device fabrication, in which the tiny block of semiconducting material is encapsulated in a supporting case that prevents physical damage and corrosion. The case, known as a "package", supports the electrical contacts which connect the device to a circuit board.

In the integrated circuit industry, the process is often referred to as packaging. Other names include semiconductor device assembly, assembly, encapsulation or sealing.

The packaging stage is followed by testing of the integrated circuit.

The term is sometimes confused with electronic packaging, which is the mounting and interconnecting of integrated circuits (and other components) onto printed-circuit boards.

Design Considerations

Electrical

The current-carrying traces that run out of the die, through the package, and into the printed circuit board (PCB) have very different electrical properties compared to on-chip signals. They require special design techniques and need much more electric power than signals confined to the chip itself. Therefore, it is important that the materials used as electrical contacts exhibit characteristics like low resistance, low capacitance and low inductance. Both the structure and materials must prioritize signal transmission properties, while minimizing any parasitic elements that could negatively affect the signal.

Controlling these characteristics is becoming increasingly important as the rest of technology begins to speed up. Packaging delays have the potential to make up almost half of a high-performance computer's delay, and this bottleneck on speed is expected to increase.

Mechanical and Thermal

The integrated circuit package is responsible for keeping the chip safe from all sorts of potential damage. The package must resist physical breakage, provide an airtight seal to keep out moisture, and also provide effective heat dissipation away from the chip. At the same time, it must have effective means of connecting to a PCB, which can change drastically depending on the package type. The materials used for the body of the package are typically either plastic or ceramic. They both can offer a high thermal conductivity and decent mechanical strength. Ceramic generally has more preferable characteristics, but is more expensive.

Increasing the surface area of the package allows for better heat transfer via convection, and some packages utilize metallic fins to enhance heat transfer even further at the cost of valuable space. Larger sizes also allow for a greater number of mechanical connections. However, these factors are balanced out by the fact that the package generally needs to be kept as small as possible.

Economic

Cost is a major limiting factor for many designs. Choices such as package material and level of precision must be balanced by the economic viability of the end product. Depending on the needs of the system, opting for lower-cost materials is often an acceptable solution to economic constraints. Typically, an inexpensive plastic package can dissipate heat up to 2W, which is sufficient for many simple applications, though a similar ceramic package can dissipate up to 50W in the same scenario. As the chips inside THE package get smaller and faster, they also tend to get hotter. As the subsequent need for more effective heat dissipation increases, the cost of packaging rises along with it. Generally, the smaller and more complex the package needs to be, the more expensive it is to manufacture.

History

Small-outline integrated circuit. This package has 14 "gull wing" leads protruding from the two long sides and a lead spacing of 0.050 inches.

The earliest integrated circuits were packaged in ceramic flat packs, which the military used for many years for their reliability and small size. Commercial circuit packaging quickly moved to the dual in-line package (DIP), first in ceramic and later in plastic. In the 1980s VLSI pin counts exceeded the practical limit for DIP packaging, leading to pin grid array (PGA) and leadless chip carrier (LCC) packages. Surface mount packaging appeared in the early 1980s and became popular in the late 1980s, using finer lead pitch with leads formed as either gull-wing or J-lead, as exemplified by small-outline integrated circuit – a carrier which occupies an area about 30 – 50% less than an equivalent DIP, with a typical thickness that is 70% less.

Early USSR made integrated circuit. The tiny block of semiconducting material (the "die"), is enclosed inside the round, metallic case (the "package").

The next big innovation was the *area array package*, which places the interconnection terminals throughout the surface area of the package, providing a greater number of connections than previous package types where only the outer perimeter is used. The first area array package was a ceramic pin grid array package. Not long after, the plastic ball grid array (BGA), another type of area array package, became one of the most commonly used packaging techniques.

In the late 1990s, plastic quad flat pack (PQFP) and thin small-outline packages (TSOP) replaced PGA packages as the most common for high pin count devices, though PGA packages are still often used for microprocessors. However, industry leaders Intel and AMD transitioned in the 2000s from PGA packages to land grid array (LGA) packages.

Ball grid array (BGA) packages have existed since the 1970s, but evolved into Flip-chip ball grid array packages (FCBGA) in the 1990s. FCBGA packages allow for much higher pin count than any existing package types. In an FCBGA package, the die is mounted upside-down (flipped) and connects to the package balls via a substrate that is similar to a printed-circuit board rather than by wires. FCBGA packages allow an array of input-output signals (called Area-I/O) to be distributed over the entire die rather than being confined to the die periphery.

Traces out of the die, through the package, and into the printed circuit board have very different electrical properties, compared to on-chip signals. They require special design techniques and need much more electric power than signals confined to the chip itself.

Recent developments consist of stacking multiple dies in single package called SiP, for *System In Package*, or three-dimensional integrated circuit. Combining multiple dies on a small substrate, often ceramic, is called an MCM, or Multi-Chip Module. The boundary between a big MCM and a small printed circuit board is sometimes blurry.

Operations

Die attachment is the step during which a die is mounted and fixed to the package or support structure (header). For high-powered applications, the die is usually eutectic bonded onto the package, using e.g. gold-tin or gold-silicon solder (for good heat conduction). For low-cost, low-powered applications, the die is often glued directly onto a substrate (such as a printed wiring board) using an epoxy adhesive.

The following operations are performed at the packaging stage, as broken down into bonding, encapsulation, and wafer bonding steps. Note that this list is not all-inclusive and not all of these operations are performed for every package, as the process is highly dependent on the package type.

- IC Bonding:
  - Wire bonding
  - Thermosonic Bonding
  - Down bonding
  - Tape-automated bonding
  - Flip chip
  - Quilt packaging
  - Tab bonding
  - Film attaching

  - Spacer attaching
- IC Encapsulation:
  - Baking
  - Plating
  - Lasermarking
  - Trim and form
- Wafer Bonding

Integrated Circuit Design

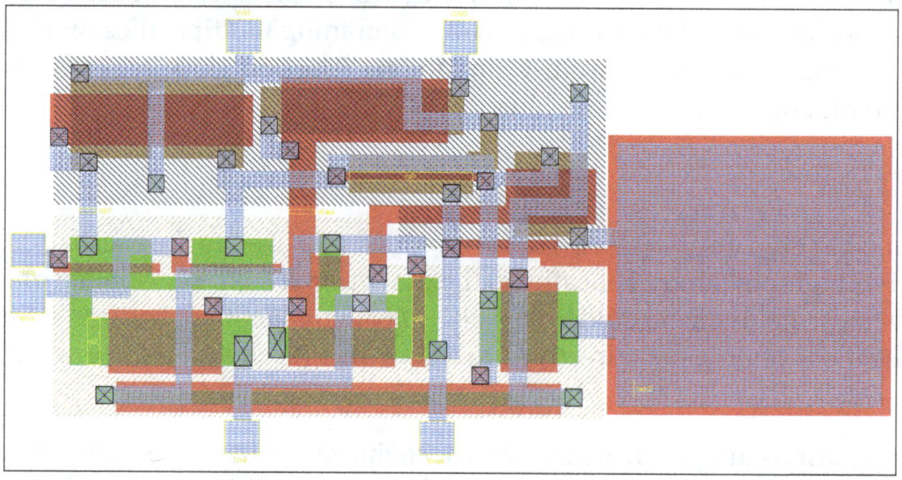

Layout view of a simple CMOS Operational Amplifier (inputs are to the left and the compensation capacitor is to the right). The metal layer is coloured blue, green and brown are N- and P-doped Si, the polysilicon is red and vias are crosses.

Integrated circuit design, or IC design, is a subset of electronics engineering, encompassing the particular logic and circuit design techniques required to design integrated circuits, or ICs. ICs consist of miniaturized electronic components built into an electrical network on a monolithic semiconductor substrate by photolithography.

IC design can be divided into the broad categories of digital and analog IC design. Digital IC design is to produce components such as microprocessors, FPGAs, memories (RAM, ROM, and flash) and digital ASICs. Digital design focuses on logical correctness, maximizing circuit density, and placing circuits so that clock and timing signals are routed efficiently. Analog IC design also has specializations in power IC design and RF IC design. Analog IC design is used in the design of op-amps, linear regulators, phase locked loops, oscillators and active filters. Analog design is more concerned with the physics of the semiconductor devices such as gain, matching, power dissipation, and resistance. Fidelity of analog signal amplification and filtering is usually critical and as a result, analog ICs use larger area active devices than digital designs and are usually less dense in circuitry.

Modern ICs are enormously complicated. An average desktop computer chip, as of 2015, has over 1 billion transistors. The rules for what can and cannot be manufactured are also extremely complex. Common IC processes of 2015 have more than 500 rules. Furthermore, since the manufacturing process itself is not completely predictable, designers must account for its statistical nature. The complexity of modern IC design, as well as market pressure to produce designs rapidly, has led to the extensive use of automated design tools in the IC design process. In short, the design of an IC using EDA software is the design, test, and verification of the instructions that the IC is to carry out.

Fundamentals

Integrated circuit design involves the creation of electronic components, such as transistors, resistors, capacitors and the metallic interconnect of these components onto a piece of semiconductor, typically silicon. A method to isolate the individual components formed in the substrate is necessary since the substrate silicon is conductive and often forms an active region of the individual components. The two common methods are p-n junction isolation and dielectric isolation. Attention must be given to power dissipation of transistors and interconnect resistances and current density of the interconnect, contacts and vias since ICs contain very tiny devices compared to discrete components, where such concerns are less of an issue. Electromigration in metallic interconnect and ESD damage to the tiny components are also of concern. Finally, the physical layout of certain circuit subblocks is typically critical, in order to achieve the desired speed of operation, to segregate noisy portions of an IC from quiet portions, to balance the effects of heat generation across the IC, or to facilitate the placement of connections to circuitry outside the IC.

Design Steps

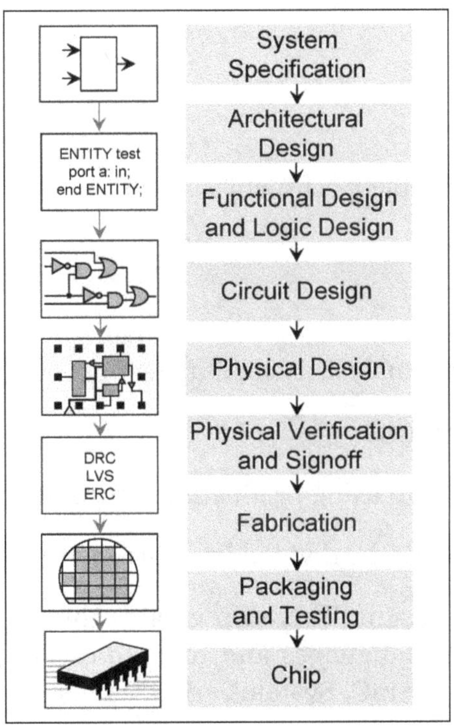

Major steps in the IC design flow.

A typical IC design cycle involves several steps:

1. Feasibility study and die size estimate.
2. Function analysis.
3. System Level Design.
4. Analogue Design, Simulation & Layout.
5. Digital Design, Simulation & Synthesis.
6. System Simulation & Verification.
7. Design For Test and Automatic test pattern generation.
8. Design for manufacturability (IC).
9. Tape-in.
10. Mask data preparation.
11. Tape-out.
12. Wafer fabrication.
13. Die test.
14. Packaging.
15. Post silicon validation and integration.
16. Device characterization.
17. Tweak (if necessary).
18. Datasheet generation Portable Document Format.
19. Ramp up.
20. Production.
21. Yield Analysis / Warranty Analysis Reliability (semiconductor).
22. Failure analysis on any returns·
23. Plan for next generation chip using production information if possible.

Roughly saying, digital IC design can be divided into three parts.

1. Electronic system-level design: This step creates the user functional specification. The user may use a variety of languages and tools to create this description. Examples include a C/C++ model, SystemC, SystemVerilog Transaction Level Models, Simulink and MATLAB.

2. RTL design: This step converts the user specification (what the user wants the chip to do) into a register transfer level (RTL) description. The RTL describes the exact behavior of the digital circuits on the chip, as well as the interconnections to inputs and outputs.
3. Physical design: This step takes the RTL, and a library of available logic gates, and creates a chip design. This involves figuring out which gates to use, defining places for them, and wiring them together.

Note that the second step, RTL design, is responsible for the chip doing the right thing. The third step, physical design, does not affect the functionality at all (if done correctly) but determines how fast the chip operates and how much it costs.

Design Process

Microarchitecture and System-level Design

The initial chip design process begins with system-level design and microarchitecture planning. Within IC design companies, management and often analytics will draft a proposal for a design team to start the design of a new chip to fit into an industry segment. Upper-level designers will meet at this stage to decide how the chip will operate functionally. This step is where an IC's functionality and design are decided. IC designers will map out the functional requirements, verification testbenches, and testing methodologies for the whole project, and will then turn the preliminary design into a system-level specification that can be simulated with simple models using languages like C++ and MATLAB and emulation tools. For pure and new designs, the system design stage is where an Instruction set and operation is planned out, and in most chips existing instruction sets are modified for newer functionality. Design at this stage is often statements such as *encodes in the MP3 format* or *implements IEEE floating-point arithmetic.* At later stages in the design process, each of these innocent looking statements expands to hundreds of pages of textual documentation.

RTL Design

Upon agreement of a system design, RTL designers then implement the functional models in a hardware description language like Verilog, SystemVerilog, or VHDL. Using digital design components like adders, shifters, and state machines as well as computer architecture concepts like pipelining, superscalar execution, and branch prediction, RTL designers will break a functional description into hardware models of components on the chip working together. Each of the simple statements described in the system design can easily turn into thousands of lines of RTL code, which is why it is extremely difficult to verify that the RTL will do the right thing in all the possible cases that the user may throw at it.

To reduce the number of functionality bugs, a separate hardware verification group will take the RTL and design testbenches and systems to check that the RTL actually is performing the same steps under many different conditions, classified as the domain of functional verification. Many techniques are used, none of them perfect but all of them useful – extensive logic simulation, formal methods, hardware emulation, lint-like code checking, code coverage, and so on.

A tiny error here can make the whole chip useless, or worse. The famous Pentium FDIV bug caused the results of a division to be wrong by at most 61 parts per million, in cases that occurred very infrequently. No one even noticed it until the chip had been in production for months. Yet Intel was forced to offer to replace, for free, every chip sold until they could fix the bug, at a cost of $475 million (US).

Physical Design

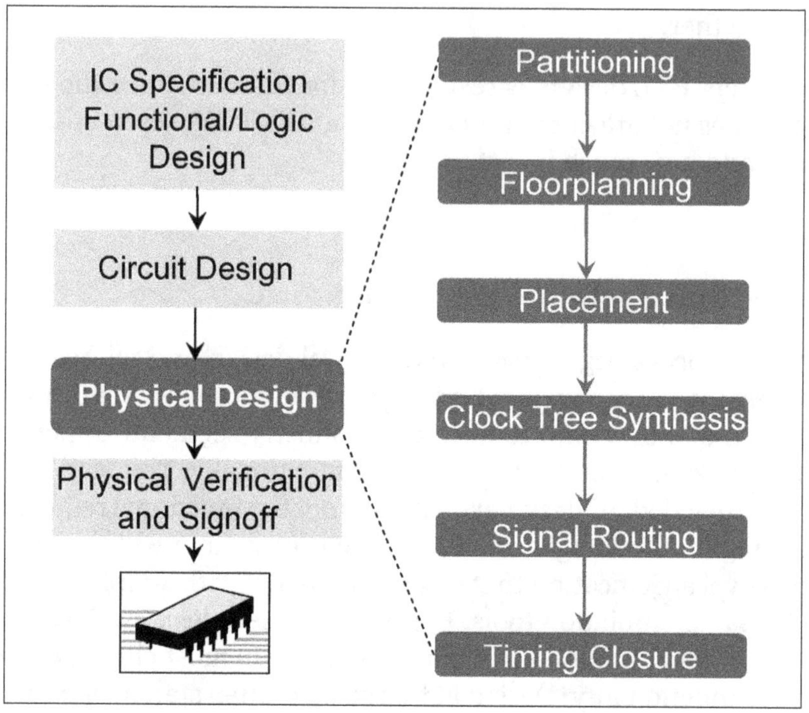

Physical design steps within the digital design flow.

RTL is only a behavioral model of the actual functionality of what the chip is supposed to operate under. It has no link to a physical aspect of how the chip would operate in real life at the materials, physics, and electrical engineering side. For this reason, the next step in the IC design process, physical design stage, is to map the RTL into actual geometric representations of all electronics devices, such as capacitors, resistors, logic gates, and transistors that will go on the chip.

The main steps of physical design are listed below. In practice there is not a straightforward progression - considerable iteration is required to ensure all objectives are met simultaneously. This is a difficult problem in its own right, called design closure.

- Logic synthesis: The RTL is mapped into a gate-level netlist in the target technology of the chip.
- Floorplanning: The RTL of the chip is assigned to gross regions of the chip, input/output (I/O) pins are assigned and large objects (arrays, cores, etc.) are placed.
- Placement: The gates in the netlist are assigned to nonoverlapping locations on the die area.
- Logic/placement refinement: Iterative logical and placement transformations to close performance and power constraints.

- Clock insertion: Clock signal wiring is (commonly, clock trees) introduced into the design.
- Routing: The wires that connect the gates in the netlist are added.
- Postwiring optimization: Performance (timing closure), noise (signal integrity), and yield (Design for manufacturability) violations are removed.
- Design for manufacturability: The design is modified, where possible, to make it as easy and efficient as possible to produce. This is achieved by adding extra vias or adding dummy metal/diffusion/poly layers wherever possible while complying to the design rules set by the foundry.
- Final checking: Since errors are expensive, time consuming and hard to spot, extensive error checking is the rule, making sure the mapping to logic was done correctly, and checking that the manufacturing rules were followed faithfully.
- Tapeout and mask generation: the design data is turned into photomasks in mask data preparation.

Analog Design

Before the advent of the microprocessor and software based design tools, analog ICs were designed using hand calculations and process kit parts. These ICs were low complexity circuits, for example, op-amps, usually involving no more than ten transistors and few connections. An iterative trial-and-error process and "overengineering" of device size was often necessary to achieve a manufacturable IC. Reuse of proven designs allowed progressively more complicated ICs to be built upon prior knowledge. When inexpensive computer processing became available in the 1970s, computer programs were written to simulate circuit designs with greater accuracy than practical by hand calculation. The first circuit simulator for analog ICs was called SPICE (Simulation Program with Integrated Circuits Emphasis). Computerized circuit simulation tools enable greater IC design complexity than hand calculations can achieve, making the design of analog ASICs practical. The computerized circuit simulators also enable mistakes to be found early in the design cycle before a physical device is fabricated. Additionally, a computerized circuit simulator can implement more sophisticated device models and circuit analysis too tedious for hand calculations, permitting Monte Carlo analysis and process sensitivity analysis to be practical. The effects of parameters such as temperature variation, doping concentration variation and statistical process variations can be simulated easily to determine if an IC design is manufacturable. Overall, computerized circuit simulation enables a higher degree of confidence that the circuit will work as expected upon manufacture.

Coping with Variability

A challenge most critical to analog IC design involves the variability of the individual devices built on the semiconductor chip. Unlike board-level circuit design which permits the designer to select devices that have each been tested and binned according to value, the device values on an IC can vary widely which are uncontrollable by the designer. For example, some IC resistors can vary ±20% and β of an integrated BJT can vary from 20 to 100. In the latest CMOS processes, β of vertical PNP transistors can even go below 1. To add to the design challenge, device properties often vary between each processed semiconductor wafer. Device properties can even vary significantly across each individual IC due to doping gradients. The underlying cause of this variability is that many semiconductor devices are highly sensitive to uncontrollable random variances in the process. Slight changes to the amount of diffusion time, uneven doping levels, etc. can have large effects on device properties.

Some design techniques used to reduce the effects of the device variation are:

- Using the ratios of resistors, which do match closely, rather than absolute resistor value.
- Using devices with matched geometrical shapes so they have matched variations.
- Making devices large so that statistical variations becomes an insignificant fraction of the overall device property.
- Segmenting large devices, such as resistors, into parts and interweaving them to cancel variations.
- Using common centroid device layout to cancel variations in devices which must match closely (such as the transistor differential pair of an op amp).

References

- Neurotechnology Group, Berlin Institute of Technology, IEEE Xplore Digital Library. "Regression Methods for Virtual Metrology of Layer Thickness in Chemical Vapor Deposition." January 17, 2014. Retrieved November 9, 2015
- R. Wayne Johnson, Mark Strickland and David Gerke, NASA Electronic Parts and Packaging Program. "3-D Packaging: A Technology Review." June 23, 2005. Retrieved July 31, 2015
- "Milestones:First Semiconductor Integrated Circuit (IC), 1958". IEEE Global History Network. IEEE. Retrieved 3 August 2011
- Bob Johnstone (1999). We were burning: Japanese entrepreneurs and the forging of the electronic age. Basic Books. pp. 47–48. ISBN 978-0-465-09118-8
- Greig, William (2007). Integrated Circuit Packaging, Assembly and Interconnections. Springer Science & Business Media. ISBN 9780387339139
- Ken Gilleo (2003). Area array packaging processes for BGA, Flip Chip, and CSP. McGraw-Hill Professional. p. 251. ISBN 0-07-142829-1
- Riley, Geroge (2009-01-30). "Flipchips: Tutorial #1". Archived from the original on January 30, 2009. Retrieved 2016-04-07

6

Diverse Fields of Applications

Electronics are applied in various areas like audio electronics, radio electronics, consumer electronics, nanoelectronics, molecular electronics, bioelectronics and nuclear electronics. The diverse applications of electronics in these areas have been thoroughly discussed in this chapter.

Audio Electronics

Audio electronics, which convert sound signals into electrical signals, process these electrical signals, and convert them back into sound.

The Basics of Sound

Before discussing the specific components of audio-related circuits and electronics, let's first take a crash course on the basic concepts of sound.

Sound is simply a type of energy vibrating through a medium (such as air or water); this energy, within a specific range of frequencies, is interpreted by the human ear as sound.

Sound is made up of three basic elements:

- Frequency: how fast the vibrations are occurring.
- Intensity: how loud the sound is.
- Timbre: the sound's quality.

The human ear can detect sound *frequencies* ranging from 20 to 20,000 Hz. However, the human ear is more sensitive to (i.e. able to discern at lowest intensity) frequencies ranging between 2,000 and 5,000 Hz. Recall that Hertz (Hz) is a unit defined as cycles per second.

A sound's *intensity* corresponds to the amount of energy associated with that sound. The decibel (dB) is used for measuring the sound's energy in a way that is relevant to how humans perceive loudness.

In the context of audio, the decibel is defined as follows:

$$dB = 10 \log_{10} (I/I_o)$$

where,

- I = the measured intensity (W/m^2).
- $I_0 = 10^{-12}$ W/m^2, which represents the lowest sound intensity detectable by the human ear.

The figure below shows a variety of sounds and their dB intensity measurement:

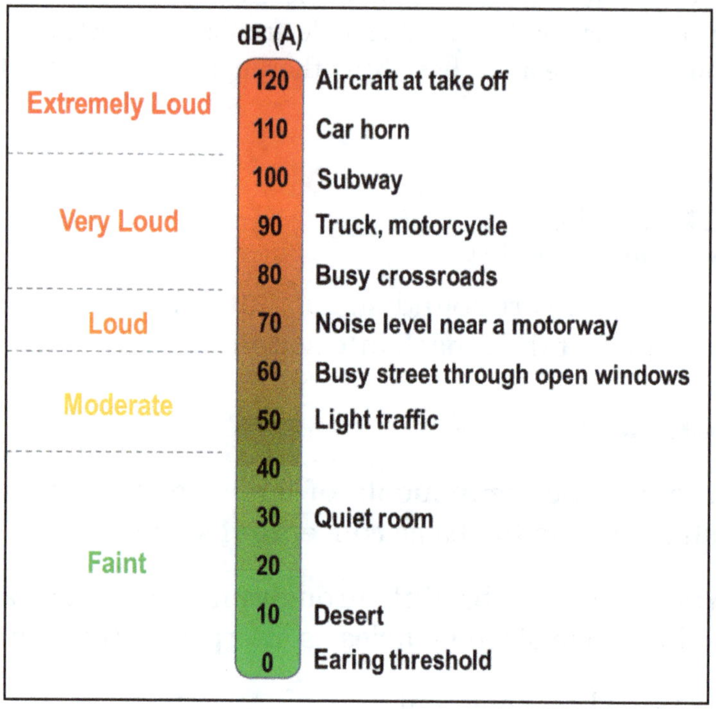

Timbre is the complex wave pattern that occurs when overtones (also referred to as harmonics) are present along with the fundamental frequency. For instance, a tuning fork is designed to produce exactly one specific frequency, called the fundamental frequency. Let's say we have a tuning fork that produces a middle C note (261.62 Hz. Since a tuning fork generates only one frequency, no other frequencies (overtones) exist.

However, if you were to hear a middle C played on a violin you would hear both the fundamental frequency (261.62 Hz) and a variety of other, albeit less intense, frequencies (overtones). The specific intensity of each overtone is largely credited for giving an instrument (or a person's voice) its unique timbre (also called tonal quality).

Microphones: Dynamic, Condenser and Electret

The microphone is responsible for converting changes in sound pressure to changes in electric current. The *intensity* of the changes in sound pressure corresponds to the AC voltage *amplitude* generated by the microphone. Likewise, the frequency of the changes in sound pressure corresponds to the frequency of the AC voltage. Obviously, if any overtones are present, they too are present in the electrical signal generated by the microphone.

Three common types of microphones are the following:

- Dynamic
- Condenser
- Electret

The Dynamic Microphone

The dynamic microphone consists of a plastic diaphragm, a voice coil and a permanent magnet.

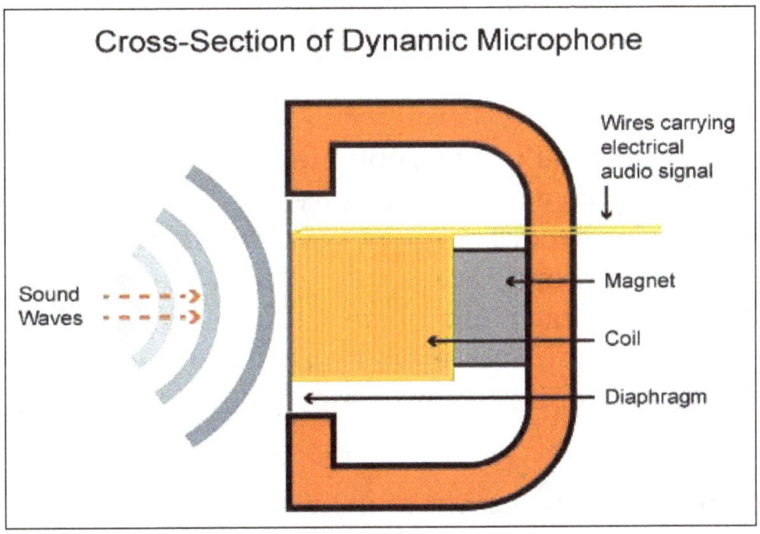

A dynamic microphone cross-section.

When varying sound pressure is applied to the diaphragm, the voice coil moves back and forth through the magnet's magnetic field, resulting in a voltage across the leads of the voice coil. Hence, the sound pressure is converted to a voltage signal.

Dynamic microphones are known for being very rugged, for being able to operate over a wide temperature range, for providing smooth and extended frequency response, and for not needing an external power source. They are widely used in applications such as public address and high-quality recording.

The Condenser Microphone

The condenser microphone uses a pair of charged plates that can be brought closer together or forced farther apart by changes in air pressure (i.e., sound). In doing so, the charged plates act like a sound-sensitive capacitor. This type of microphone works in conjunction with a low-noise, high-impedance amplifier.

Condenser microphones are known for providing crisp, low-noise sound and are used for producing high-quality sound recordings.

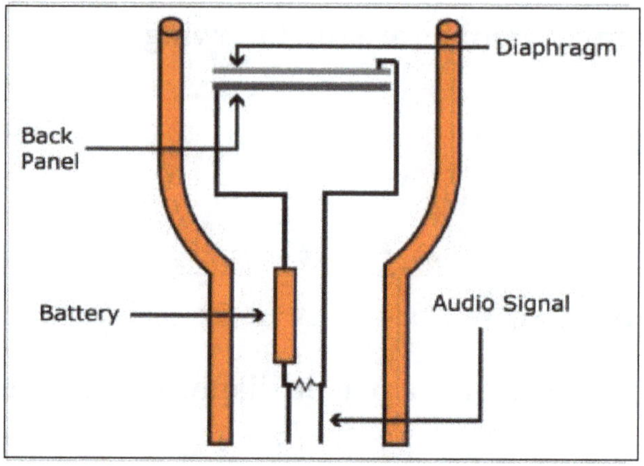

A condenser microphone's basic components.

The Electret Microphone

An *electret* microphone is actually a microphone and a JFET in a single package. The gate of the JFET provides a very high impedance and thus ensures that the charge on the electret element remains fixed. The charge must remain fixed to ensure that changes in capacitance (caused by sound waves) result in voltage changes. The JFET also converts the varying (gate) voltage into varying (drain-source) current. While the electret element itself does not require power (because its permanently charged) the overall electret microphone device requires a power supply because the integrated JFET needs bias current.

Older electret microphones were known to deliver poor performance, but modern devices can compete with condenser microphones.

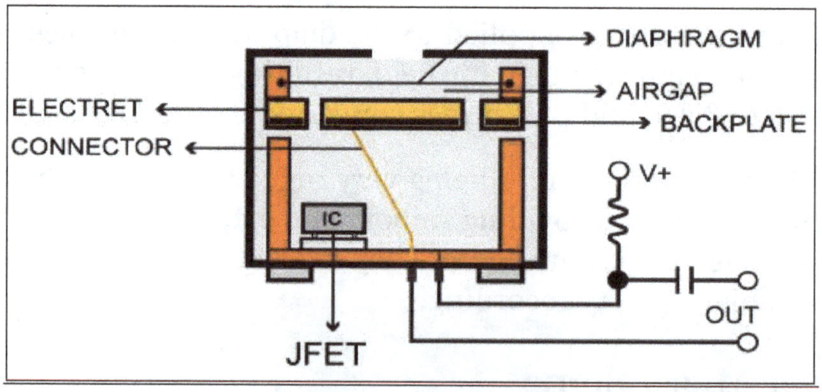

An electret microphone.

Types of Speakers

Microphones convert sound signals into electrical signals. Speakers do the exact opposite they convert electrical signals into audible (sound) signals.

The dynamic speaker, which uses the same operating principles as the dynamic microphone, is the most popular speaker used today. When a varying electrical signal (current) is channeled through a moveable coil (voice coil) surrounded by a magnet, the coil moves back and forth. Cones (often made of paper) attached to the moving coil then produce variations in air pressure (i.e. sound waves) that correspond to the electrical signal driven through the speaker coil.

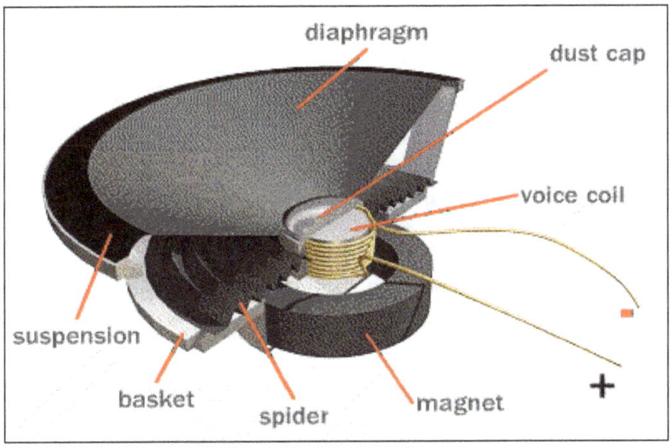

Parts of a dynamic speaker.

Every speaker has a nominal resistance—or, more accurately, an impedance—that represents the average resistance between the speaker leads. Thus, speakers behave similarly to resistors when placed in series and in parallel.

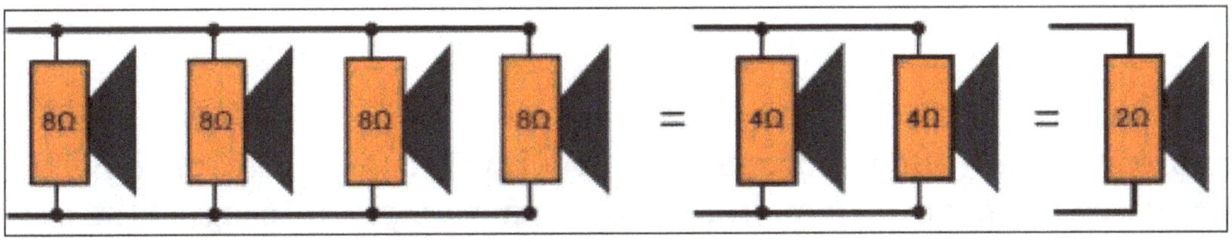

Speakers in parallel behave the same as resistors in parallel.

Frequency response, an important speaker characteristic, represents the frequency range over which a speaker can successfully propagate sound waves.

Speakers can be categorized according to the frequency range that they are designed for:

- Woofers: speakers designed specifically for low frequencies (less than 200 Hz).
- Midrange: speakers designed to accommodate frequencies ranging from 500 Hz to 3000 Hz.
- Tweeter: a dedicated speaker type specifically designed to handle frequencies above those of midrange speakers.
- Full-range speakers: capable of handling frequencies ranging from 100 Hz to 15,000 Hz.

Full-range speakers have inferior sound qualities when compared to a sound system consisting of woofers, midrange speakers, and tweeters. Such a three-way speaker system is illustrated below:

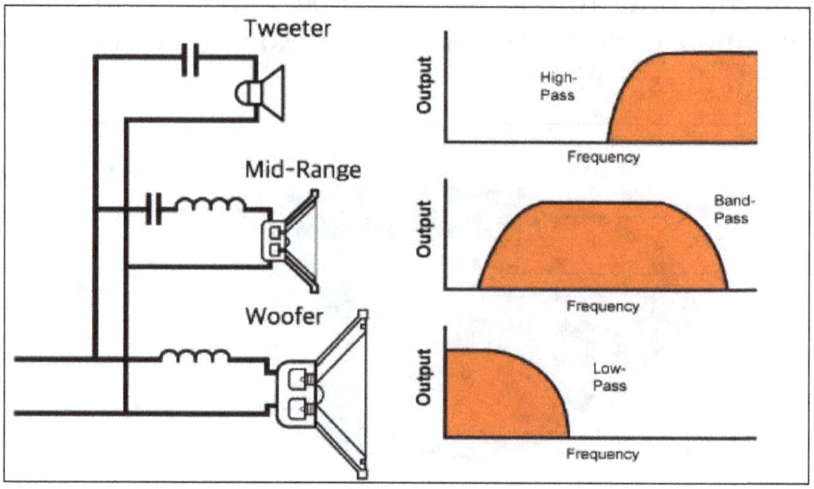

A typical three-way speaker system consisting of a tweeter, a midrange speaker, and a woofer.

Amplifiers 101

Audio amplifier have three types of classification:

- Preamplifier
- Low-power amplifier
- Power amplifier

An audio preamplifier (often shortened to "pre-amp") is an electronic device that amplifies a very weak signal from a microphone, as an example, into signals strong enough to manipulate. Pre-amps are often simple, fixed-gain amplifiers designed specifically for low-noise performance.

A *low-power amplifier* is often used to manipulate signals including such aspects as volume and frequency equalization. This type of amplifier generally focuses on changing the character of the signal in desired ways while introducing as little unwanted distortion as possible and may provide little to no actual power amplification.

An audio *power amplifier* ("power amp") is used to increase the signal power so as to drive a load, such as output speakers. Similar to pre-amps, power amps are often fixed gain (in terms of signal amplitude) so that designers can focus on high-power gain and the power handling challenges that typically result. In simple audio systems where high power and high fidelity are not critical factors, a single-amplifier circuit may perform all of these functions, and, in fact, specially designed operational amplifiers, such as the LM-386 Low Voltage Audio Power Amplifier, are often used this way. A Class D audio amplifier is a switching or PWM amplifier where the switches are either fully ON or fully OFF, which significantly reduces the power losses. Class D amplifiers take on many different forms, including some that have digital inputs while others have analog inputs.

Radio Electronics

Radio Electronics is a term encompassing a wide range of areas in science and engineering that deal with the problems of transmission, reception, and processing of information by means of electromagnetic waves. The term came into use during the 1950's and to a certain extent is conventional.

Radio electronics encompasses radio engineering and electronics, as well as a number of new areas that have emerged as a result of development and differentiation. Such areas include quantum electronics, optoelectronics, semiconductor electronics, microelectronics, infrared technology, cryoelectronics, acoustical electronics, and chemotronics. Radio-electronics is closely associated with radio physics, solid-state physics, optics, and mechanics, as well as with electrical engineering, automation, and engineering cybernetics. Radio-electronic apparatus often serve as one of the parts inautomatic control systems, such as guidance systems that govern the flight of rockets and space craft. Conversely, automatic control systems may be used in such apparatus, for example, in automatic frequency control or target tracking.

Radio electronics is also associated with computer engineering, since this branch of engineering uses electronic devices inthe processing of information, as in noise suppression or reduction of data to a pre-determined form. Frequency bands that are used in radio electronics overlap those used in electroacoustics. Mathematical studies are widely used in radioelectronics for the analysis and synthesis (design) of radio-engineering circuits and devices and for the determination ofoptimum design and parameters.

Consumer Electronics

Consumer electronics includes a broad set of electronic devices that provide one or more functionalities in a home or for a particular individual. The term initially referred to electronic devices that were installed or used specifically inside a home/house. However, they also now incorporate mobile and computing devices, which can easily be carried by an individual outside the home, such as a cell phone or a tablet PC.

Consumer electronics include items, such as:

- Televisions

- DVD players
- Refrigerators
- Washing machines
- Computers
- Laptops
- Tablets

Nanoelectronics

Nanoelectronics is based on the application of nanotechnology in the field of electronics and electronic components. Although the term Nanoelectronics may generally mean all the electronic components, special attention is given in the case of transistors. These transistors have a size lesser than 100 nanometres. Visibly, they are very small that separate studies have to be made for knowing the quantum mechanical properties and inter-atomic design. As a result, though the transistors appear in the nanometre range, they are designed through nanotechnology. Their design is also very much different from the traditional transistors and usually falls in the category of one-dimensional nanotubes/nanowires, hybrid molecular electronics, or advanced molecular electronics.

This technology is said to be the next future, but its practicality is near to impossible even now that they may be difficult to emerge soon.

Basic Concept of Nanoelectronics

Although a nanoelectronic device can be made fully functional, the work load it can do is restricted to its size. The basic principle is that the power of a machine will increase according to the increase in volume, but the amount of friction that the machine's bearings hold will depend on the surface area of the machine.

For the small size of the nanoelectronic device cannot be used for the moving of heavy load like a mechanical device. If such a task is tried, it will fail as the available power will be easily overcome by the frictional forces. So, it is sure that these devices have limitations in real world applications.

Different Approaches to Nanoelectronics

- Nanofabrication

This method is used to design arrays or layers of nanoelectronic device to work for a single operation. Nanoelectro-mechanical systems are also a part of nanofabrication.

- Nanomaterials electronics

In Nanoelectronics, the transistors are packed as arrays on to a single chip. Thus they remain in a uniform manner and symmetrical in nature. Thus they are known to have a more speedy movement of electrons in the material. The dielectric constant of the device also increases and the electron or hole characteristics also become symmetrical in nature.

Besides transistors, nanoelectronic devices play a role in data storage (memory). Here, *spintronics* – the study and exploitation in solid-state devices of electron spin and its associated magnetic moment, along with electric charge is already an established technology.

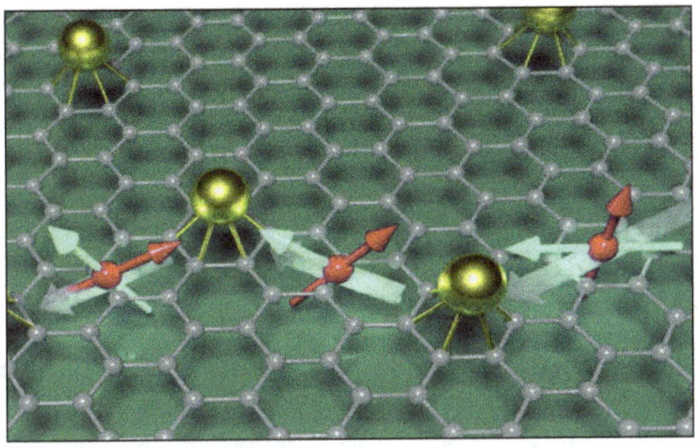

Illustration of electron spin in a graphene lattice.

Spintronics also plays a role in new technologies that exploit quantum behavior for computing.

Applications of Nanoelectronics

Some of the devices that have been developed with the help of Nanoelectronics and its future applications are listed below:

- Nanoradio
- Nanocomputers

The conventional computers with a big processor will be replaced with Nanocomputers with nanoprocessors that will have higher performance and speed than the conventional computers. Researchers are performing various experiments on by using nanolithographic methods to design better nanoprocessors. Experiments are also taking place by replacing the CMOS components in conventional processors with nanowires. The FET's in the computers are replaced by carbon nanotubes.

- Energy production

The devices using Nanoelectronics technology also includes solar cells that are highly efficient and cheaper than the conventional ones. If such efficient solar energy can be created it would be a revolution to the global energy needs.

Using the technology, researchers are developing a generator for energy production in vivo called bio-nano generators. Basically, the generator is an electrochemical device which is designed in nanoscale size. It works like a fuel cell which generates the power by absorbing the blood glucose in a living body. The glucose will be separated from the body with the help of an enzyme. This enzyme separates the glucose from the electrons and makes them useful for generating power.

The power generated through such a device will be only a few watts as the body itself needs some glucose for its normal functioning. This small power can be used to power up devices placed inside the body like pacemakers or sugar-fed nanorobots.

Molecular Electronics

Molecular electronics is an interdisciplinary field that involves the use of molecules as fundamental electronic components such as wires, transistors, memory cells, and logic elements. The field, sometimes called "moletronics" is currently attracting attention from a broad cross-section of the scientific community in response to both fascination with the fundamental scientific challenges associated with the measurement of the electrical properties of molecules, and manipulation of electronic phenomena via molecular processes, and growing concerns over the technological challenges and ultimate limits facing solid state semiconductor technology.

Before entering discussion, a distinction needs to be made between molecular materials for electronics and single molecule electronics. The use of organic materials for electronic applications in which the bulk electronic or optoelectronic response arises from ensembles of several millions of molecules, and for which properties are measured or observed on the macroscopic level, is at a mature stage of development and application. The most obvious examples of molecular materials readily available in the electronics mass market are the use of liquid crystals and organic light emitting diodes (OLEDs) in flat video displays. This sector of the electronics industry continues its steady development driven by the promises of transparent, flexible, non-toxic, printable electronics, with improved consumer appeal and reduced fabrication costs.

However, the field of single molecule electronics, which is the subject of this review, can be summarised as the use of individual molecules to mimic functional elements in electronic devices. The use of molecules as building blocks to give rise to a more complex system is typically referred to as a the "bottom-up" approach, as opposed to the conventional "top-down" lithographic techniques that are employed to etch small features into single-crystal silicon wafers. Despite, or perhaps because of, the great challenges the field of single molecule electronics presents, the enormous benefits of a molecule-based "bottom-up" approach motivates the scientific community to keep moving the field forward. For example, a typical computer microchip nowadays contains of the order of

109 transistors in areas of about 3 cm 2 with all its components working in perfect harmony and capable of changing state in response to an input signal at an astonishing rate and for an almost unimaginable number of cycles. The enormous challenge for single molecule electronics lies in the development of molecular systems that perform in some way better than, or perhaps provide an alternative function to, these extraordinary solid-state devices.

Bioelectronics

Bioelectronics is the application of the principles of electronics to biology and medicine. Hollywood has expanded the horizons with its imagination, and movies like Star Wars, Star Trek, iRobot, Minority Report and many other sci-fi movies have shown us what modern technology is capable of. The potential of bioelectronics is also shown in these movies. Artificial limbs, humanoids, the various sensors that are attached to the body, etc. are the applications of bioelectronics.

Bioelectronics encompasses a range of topics at the interface of biology and electronics. One aspect of bioelectronics is the application of electronics to problems in biology, medicine, and security. This includes electronics for both detection and characterization of biological materials, such as on the cellular and subcellular level. Another aspect of bioelectronics is using biological systems in electronic applications (e.g., processing novel electronic components from DNA, nerves, or cells). Bioelectronics also focuses on physically interfacing electronic devices with biological systems (e.g., brain-machine, cell-electrode, or protein-electrode).

Applications in this area include assistive technologies for individuals with brain-related disease or injury, such as paralysis, artificial retinas, and new technologies for protein structure-function measurements.

Examples of Technical Areas of Opportunity for Bioelectronics

The overarching technical drivers for bioelectronics are the constant advances in semiconductor technology coupled with advances in surface chemistry and their application to life sciences research. Much of the work that has been done in the bioelectronics field since the 1990s has been focused on creating better biosensors by integrating biomolecules with semiconductors. For decades, semiconductor technology has advanced at an exponential rate and is described by Moore's law, which states that the number of features in a given area of substrate doubles every 18–24 months. A result of Moore's law is that the computing power and capabilities increase with each generation of device while the cost per function decreases.

As Moore's law has progressed, so have the number of semiconductor applications in the life sciences. In particular, significant effort has gone into developing surface chemistries that can be used to attach biological molecules to semiconductor substrates. Two examples are biosensors that use enzymes covalently linked to electrode surfaces and DNA recognition based on surface-bound DNAfunctionalized polypyrrole molecules. Although advances have been made in binding biological molecules to substrates, no truly biomimetic synthetic surfaces yet exist (e.g., for use in implants).

The following are examples of technical areas or opportunities that will impact bioelectronics:

Real-Time and Massively Parallel Molecular and Cellular Characterization for Systems Biology

The nascent field of systems biology using systems engineering approaches to analyze cellular function is driving the development of new technology that can monitor multiple aspects of cellular behavior over many timepoints. Systems biology embodies a new perspective from which to view biological systems and knowledge culled from its approaches could lead to advances in medicine and security. However, significant investment is needed to develop tools and associated standards and metrology that can characterize and continuously monitor the states of cells at sub-cellular resolutions. Advances in micro- and nano-fluidic systems that can monitor many cell populations in parallel are beginning to address this need.

Biologically-based Sensors and Fabrication

Another driver of bioelectronics is that cells and their components can be used as biological transducers for measurements or as components in building novel materials or circuits. In other words, components of cells have been used for novel applications outside of their originally observed purpose. Cells are sensitive to many environmental cues. For example, identifying the response of a cell to a known toxin could allow it to be used as a "canary in a coal mine" to detect toxic substances in captured air samples. Biomolecules, in particular antibodies can also be used as transducers, via their exquisite specificity for complementary molecules. Coupling antibodies with emerging nano-scale technologies could result in ultra-sensitive detection methods. Bio-inspired fabrication shows promise for constructing nanoscale assemblies, which could lead to significant advances in sensor design and materials technology.

Protection and Restoration of Health

The development of new technology that can protect lives (e.g., by preventing bioterrorism) and restore health (e.g., by developing novel implantable therapeutic devices) is also driving the development of bioelectronics. National security demands new technology for monitoring chemical or biological threats, and advances in lab-on-a-chip technology have been developed to address this need. Further miniaturization of these technologies will likely result in smaller units with increasing capabilities. Advances in miniaturization and power transmission, storage, and generation have allowed implantable medical devices to emerge. The artificial retina, which is a first step toward restoring sight in people with degenerative diseases of the retina, is representative of these advancements. Implantable drug delivery devices, based on micro-electro-mechanical systems (MEMS), are beginning to emerge, and the next generation of drug delivery devices are likely to be "smart" and to have embedded transceivers (i.e., able to sense and respond to their environment and transmit and receive data) instead of passively delivering drugs at pre-defined intervals.

Applications of Bioelectronics

Pacemakers: Also called artificial pacemakers, have been a boon to patients by regulating their heart beats. Initially they weren't monitored electronically but eventually due to the development in the field of bioelectronics and biotechnology they work perfectly, have given life to thousands across the world, and make complex problems concerned with the heart simple.

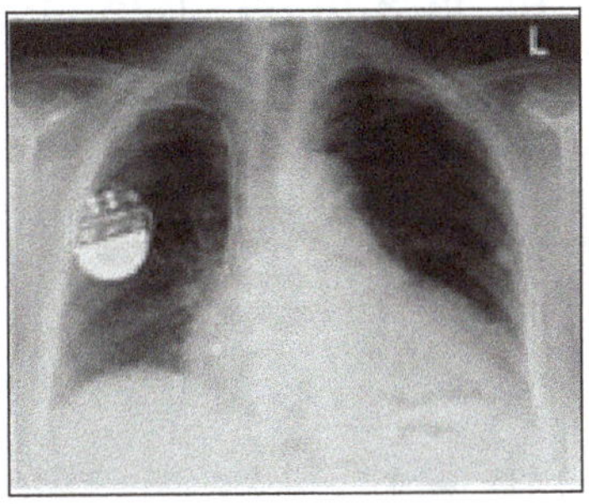

Artificial Limbs: For people who have lost their limbs in accidents or some natural cause prosthetics are used. Bioelectronics is used to power them, control them, modify and manipulate their structure as per the needs. Thus the affected ones are given a second chance to live thanks to bioelectronics and its principles.

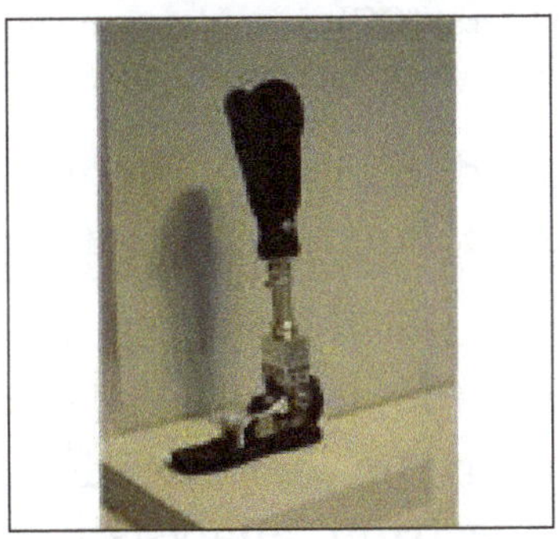

Blood Glucose Meter: Patients with diabetes need to keep their blood glucose level in check. Various blood glucose meters are available in the market that help patients to check their blood glucose in their own home instead of the olden days method of going to some centers and getting the blood glucose level checked. They give accurate readings with just a couple of drops of blood and there is virtually no pain in the entire process. Thus the patient can keep check of his or her blood glucose level and this will keep the patient out of danger.

Nuclear Electronics

Nuclear Electronics is the aggregate of methods of nuclear physics in which electronic instruments are used to receive, convert, and process information from nuclear radiation detectors. In addition to applications in nuclear physics and elementary particle physics,such methods are used wherever work involves ionizing radiation—in chemistry, medicine, space research and elsewhere.

Because of the short duration and (usually) high frequency of the processes in question as well as the presence ofbackground noise, instruments used in nuclear electronics must have a high time resolution ($\sim 10^{-9}$ sec). Because of the necessity of simultaneous measurement of a large number of parameters (signal amplitude and arrival time, the coordinates of the detection point, and so on), nuclear electronics was the first field for which analogue-digital.

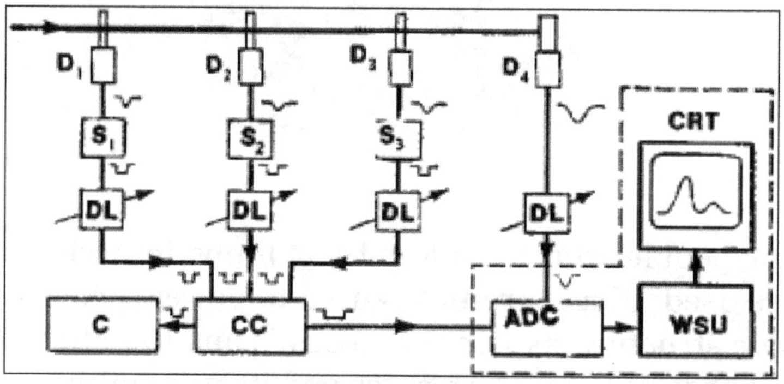

Figure: Diagram of a charged-particle spectrometer: (D_1), (D_2), (D_3), and (D_4) detectors, (S_1), (S_2), and (S_3) pulse shapers, (CC)coincidence circuit, (DL) delay lines, (ADC) analogue-to-digital converter, (WSU) working storage unit, (CRT) cathode-ray tube, (C) counter.

Conversion circuits were developed, digital methods of data accumulation were used, multichannel and multidimensionalanalyses were conducted, and computers were used.

In the registration of particles (or quanta) the task of nuclear electronics is to count pulses received from a detector. In the identification of the type of radiation or the investigation of the radiation's spectrum, analysis is made of pulse wave form, pulse amplitude, or the relative delay between pulses. In the investigation of the spatial distribution of radiation, the numberof detectors that have fired is recorded or the coordinate of the detection point may be determined directly.

The principal elements of the devices used in nuclear electronics include coincidence circuits, anti-coincidence circuits, amplitude discriminators, linear gating circuits and adders, multichannel time and amplitude analyzers and various devices that give a plot of information from coordinate detectors such as spark chambers and proportional ionization chambers. A complete listing would include hundreds of items.

A device for registering particles contains a detector, amplifier, signal converter and registering device. The converter transforms the detector signal into a standard pulse or converts the amplitude or arrival time of the signal to digital code. Pulse counters, memory devices or computers may be used to record measurement results; chart recorders and camerasare less often used.

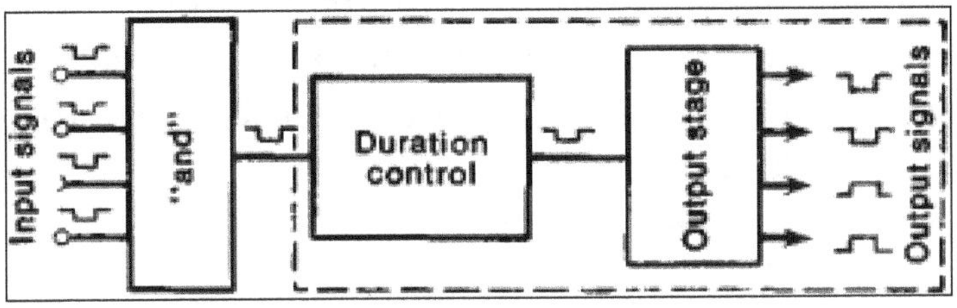

Figure: Coincidence circuit.

A simplified system for studying radiation spectra is shown in figure above. A charged particle intersects detectors (D_1), (D_2) and (D_3) and is stopped in detector (D_4). The signals from detectors (D_1), (D_2) and (D_3) are fed through pulse shapers (S_1), (S_2) and (S_3) to a coincidence circuit (CC), which selects events whose signals arrive at its inputs simultaneously. Simultaneity of pulse arrival is ensured by delay lines (DL). The coincidence circuit produces a signal that allows conversion of the pulse under investigation from detector (D_4). The converted pulse is fed as a digital code from an analogue-to-digital converter (ADC) to a working storage unit (WSU) or a computer. The measured amplitude spectrum is displayed on the screen of acathode-ray tube (CRT). This part of the system, enclosed within the broken line, is a multi-channel amplitude analyzer. The count rate at the output of the coincidence circuit, as read out by a counter (C), shows the number of registered events. The time selection of signals is handled by coincidence circuits that operate from pulses with a specified duration and amplitude. The coincidence circuits execute the logical operation "and" (logical multiplication); that is a signal appears at the output ofthe circuit only when the pulses at all inputs are of a specified level, called the unit level. If a signal with inverted polarity isapplied to one of the inputs of a coincidence circuit, the circuit becomes an anti-coincidence circuit. Modern coincidence and anti-coincidence circuits use standard integrated circuitry.

Amplitude selection is handled by discriminators based on a Schmitt trigger or tunnel diodes. The discriminators produce a standard output pulse only if the voltage (or current) at the input exceeds a specified level. Comparators are often used for amplitude discrimination. The evolution of coincidence circuits and amplitude discriminators is typical of other nuclearelectronics instruments as well. In addition to units that execute a single logical operation ("and," "or" and so on), universal multi-operation devices have been developed in which the logical operation can be assigned externally. The development of such devices has been made possible by the introduction of computers in nuclear electronics.

Computer technology has also facilitated the development of automated equipment with program control of parameters: the computer controls the operating thresholds of gating circuits, time resolution, signal delay, the logic of event selection, and the mode of operation of the measuring system. Microprocessors and specialized processors for pattern recognition and for accumulation and preliminary processing of measurement results have also been introduced into practice in physicale xperiments. Experimental data are accumulated in the computer with subsequent transcription on to magnetic tape. The results of preliminary processing are displayed on the screen of a cathode-ray tube, enabling the operator to intervene during the measurement process. The computer controls various actuating devices motors that move the detectors or target, relays, signal commutators and the like.

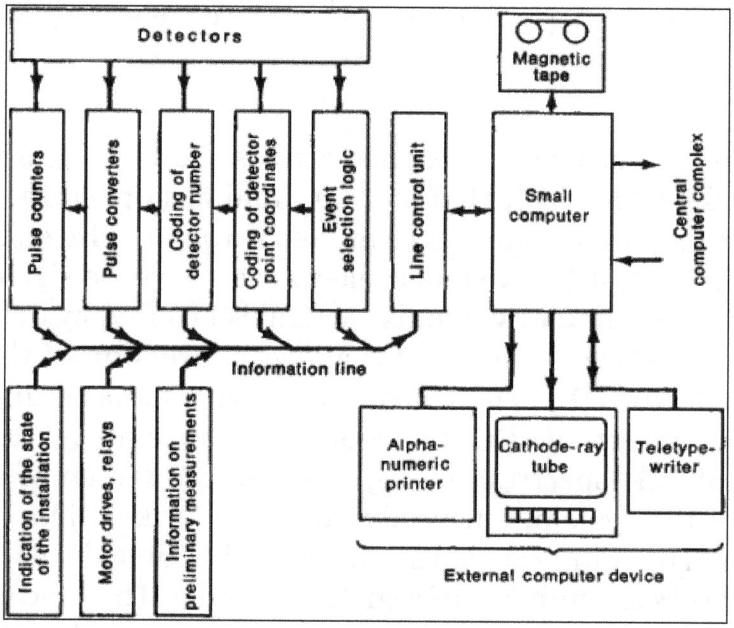

Figure: System for accumulating and processing information in a nuclear physicsexperiment.

References

- Molecular-Electronics-Revised-11601647: researchrepository.uwa.edu.au, Retrieved 05 April 2020
- Overcoming-disability-with-bioelectronics-73831, consumer-appliances-electronics: brighthubengineering.com, Retrieved 25 June 2020
- Nanoelectronics: circuitstoday.com, Retrieved 05 July 2020
- Introduction-audio-electronics-sound-microphones-speakers-amplifiers: allaboutcircuits.com, Retrieved 05 March 2020
- Consumer-electronics-ce-757: techopedia.com, Retrieved 25 April 2020

Permissions

All chapters in this book are published with permission under the Creative Commons Attribution Share Alike License or equivalent. Every chapter published in this book has been scrutinized by our experts. Their significance has been extensively debated. The topics covered herein carry significant information for a comprehensive understanding. They may even be implemented as practical applications or may be referred to as a beginning point for further studies.

We would like to thank the editorial team for lending their expertise to make the book truly unique. They have played a crucial role in the development of this book. Without their invaluable contributions this book wouldn't have been possible. They have made vital efforts to compile up to date information on the varied aspects of this subject to make this book a valuable addition to the collection of many professionals and students.

This book was conceptualized with the vision of imparting up-to-date and integrated information in this field. To ensure the same, a matchless editorial board was set up. Every individual on the board went through rigorous rounds of assessment to prove their worth. After which they invested a large part of their time researching and compiling the most relevant data for our readers.

The editorial board has been involved in producing this book since its inception. They have spent rigorous hours researching and exploring the diverse topics which have resulted in the successful publishing of this book. They have passed on their knowledge of decades through this book. To expedite this challenging task, the publisher supported the team at every step. A small team of assistant editors was also appointed to further simplify the editing procedure and attain best results for the readers.

Apart from the editorial board, the designing team has also invested a significant amount of their time in understanding the subject and creating the most relevant covers. They scrutinized every image to scout for the most suitable representation of the subject and create an appropriate cover for the book.

The publishing team has been an ardent support to the editorial, designing and production team. Their endless efforts to recruit the best for this project, has resulted in the accomplishment of this book. They are a veteran in the field of academics and their pool of knowledge is as vast as their experience in printing. Their expertise and guidance has proved useful at every step. Their uncompromising quality standards have made this book an exceptional effort. Their encouragement from time to time has been an inspiration for everyone.

The publisher and the editorial board hope that this book will prove to be a valuable piece of knowledge for students, practitioners and scholars across the globe.

Index

www.ingramcontent.com/pod-product-compliance
Lightning Source LLC
LaVergne TN
LVHW080715210926
848485LV00003B/469